The Open
University

S342
Science: a third level course

PHYSICAL CHEMISTRY

PRINCIPLES OF CHEMICAL CHANGE

TOPIC STUDY 1
THE THREAT TO STRATOSPHERIC OZONE

THE S342 COURSE TEAM

CHAIR AND GENERAL EDITOR

Kiki Warr

AUTHORS

Keith Bolton (Block 8; Topic Study 3)

Angela Chapman (Block 4)

Eleanor Crabb (Block 5; Topic Study 2)

Charlie Harding (Block 6; Topic Study 2)

Clive McKee (Block 6)

Michael Mortimer (Blocks 2, 3 and 5)

Kiki Warr (Blocks 1, 4, 7 and 8; Topic Study 1)

Ruth Williams (Block 3)

Other authors whose previous S342 contribution has been
of considerable value in the preparation of this Course

Lesley Smart (Block 6)

Peter Taylor (Blocks 3 and 4)

Dr J. M. West (University of Sheffield, Topic Study 3)

COURSE MANAGER

Mike Bullivant

EDITORS

Ian Nuttall

Dick Sharp

BBC

David Jackson

Ian Thomas

GRAPHIC DESIGN

Debbie Crouch (Designer)

Howard Taylor (Graphic Artist)

COURSE READER

Dr Clive McKee

COURSE ASSESSOR

Professor P. G. Ashmore (original course)

Dr David Whan (revised course)

SECRETARIAL SUPPORT

Debbie Gingell (Course Secretary)

Jenny Burrage

Margaret Careford

Shirley Foster

The Open University, Walton Hall, Milton Keynes, MK7 6AA

Copyright © 1996 The Open University. First published 1996

Edited, designed and typeset by The Open University.

Printed in the United Kingdom by Martins the Printers Ltd, Berwick upon Tweed

ISBN 0 7492 51905

This text forms part of an Open University Third Level Course. If you would like a copy of Studying with The
Open University, please write to the Central Enquiry Service, PO Box 200, The Open University, Walton Hall,
Milton Keynes, MK7 6YZ. If you have not enrolled on the Course and would like to buy this or other Open
University material, please write to Open University Educational Enterprises Ltd, 12 Cofferidge Close, Stony
Stratford, Milton Keynes, MK11 1BY, United Kingdom.

s342TS1i1.2

CONTENTS

1 INTRODUCTION 5

2 THE ATMOSPHERE:
A BRIEF LOOK AT ITS STRUCTURE AND COMPOSITION 7
2.1 Ozone in the atmosphere 9

3 THE STRATOSPHERIC OZONE CYCLE 10
3.1 A simple photochemical model: 'oxygen-only' chemistry 12
3.2 A kinetic analysis of the Chapman mechanism 13
3.3 The influence of trace species: catalytic cycles 17
3.4 A closer look at the catalytic 'families' 19
3.5 Source gases – natural and unnatural 21
3.6 Summary of Sections 2 and 3 25

4 MODELLING STUDIES:
A HISTORICAL PERSPECTIVE 28
4.1 The global distribution of stratospheric ozone:
the influence of transport processes 29
4.2 Formulating a model 30
4.3 Model predictions of global ozone depletion:
changing perceptions 31
4.4 Summary of Section 4 36

5 NEW SCIENCE: NEW URGENCY 37
5.1 Antarctica: the ozone hole 38
5.2 Global ozone trends 43
5.3 Drawing the threads together 48

6 WHAT MIGHT OZONE LOSS MEAN IN PRACTICE? 50
6.1 Biological effects of enhanced uv radiation at the surface 50
6.2 Effects linked to ozone's role in the climate system 53
6.3 Summary of Section 6 54

7 STRATEGIES FOR PROTECTING THE
OZONE LAYER 55
7.1 Forging international agreement 55
7.2 Science and the review process:
the Montreal Protocol in evolution 56
7.3 Looking to the future 61

OBJECTIVES FOR TOPIC STUDY 1 63

SAQ ANSWERS AND COMMENTS 64

ANSWERS TO EXERCISES 69

ACKNOWLEDGEMENTS 72

1 INTRODUCTION

The Montreal Protocol was signed on 16 September 1987. It was the first international treaty to restrict the use of substances deemed damaging to the *global* environment – specifically, to the stratospheric ozone layer. Located several miles above the surface, the ozone layer acts as a filter, protecting life on Earth from the full intensity of the Sun's ultraviolet radiation.

The Protocol embodied international concern about the widespread, and potentially serious, consequences of damage to this vital shield. It effectively sounded the death knell for the chlorofluorocarbons (CFCs) – an extremely useful group of synthetic halocarbons. CFCs first came on stream in the 1930s, as the 'ideal' – stable, non-flammable, non-toxic – substitutes for the decidedly noxious chemicals then used as refrigerants. They are also cheap to produce, and have physical properties well suited to a wide range of other applications – as propellants in aerosol spray-cans, as standard ingredients in rigid and flexible plastic-foam materials, and as solvents and cleaning fluids in many speciality areas (notably, the electronics industry). The proliferating uses of CFCs saw annual production of these compounds soar to over 800 000 tonnes by the mid-1970s.

Unless steps are taken to prevent it, virtually all uses of CFCs result in the compounds eventually being released to the atmosphere. The legacy of some 60 years of uncontrolled release is manifest in the massive loss of stratospheric ozone over Antarctica that occurs with the return of sunlight each southern spring – the so-called 'ozone hole', first reported in 1985. By 1988, there was irrefutable evidence linking this seasonal phenomenon with the presence in the stratosphere of chlorine atoms, largely derived from the breakdown of CFCs. And there is now growing evidence of stratospheric ozone depletion at other latitudes – and in both hemispheres – as well.

1985

Antarctic stratosphere is losing ozone

John Gribbin

THE BRITISH Antarctic Survey has discovered significant seasonal reductions in the ozone content of the stratosphere above Antarctica. The finding should help determine whether the world's protective layer of ozone is declining.

It is now more than 10 years since American scientists Sherry Rowland and Mario Molina first warned that chlorofluorocarbons (CFCs) might cause a breakdown of ozone in the stratosphere, the layer of the atmosphere from about 6-15 kilometres to about 50 kilometres above the Earth. CFC gases are very long-lived and stable, which is why they were favoured by the

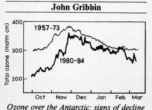

Ozone over the Antarctic: signs of decline

of the years 1957-1973 and a similar composite of the years 1980-84. The more recent measurements, published in last week's *Nature*, show less ozone present in all seasons measured, but the difference is most significant for the spring, that is September to November.

It is impossible to be sure just why this change has occurred. But the Cambridge team suggests that nitrogen oxides are locked up as N_2O_5 during the long winter night. Then, in the spring, while the air is still cold, the long slanting rays of the Sun begin to stimulate photochemical reactions. These produce oxides such as NO_3

Prompted by these disturbing developments, the parties to the Protocol (148 by late 1994) have since agreed a tightening of both the scope and the timing of the controls set out in the 1987 agreement. As a result, production of CFCs (and several other synthetic halocarbons) should have been phased out completely by the time you read this text. This action was the culmination of a long-running debate about CFCs, initiated in 1974 by F. Sherwood Rowland and Mario Molina. Their warning about the possible dangers of continued release of CFCs was prompted by the observed accumulation of these compounds in the atmosphere, and backed up by theoretical calculations based on general physical and chemical principles. It came as an economic, as well as an environmental, bombshell. Not surprisingly, the topic rapidly blossomed into one of the more controversial scientific issues of recent times – a feature evident in the extracts collected overleaf.

The perceived threat to stratospheric ozone has, in recent decades, motivated a massive programme of research devoted to the various facets of the problem. This ongoing programme has produced an increasingly detailed understanding of the processes that maintain the ozone layer – and of the way in which human activities are interfering with the natural control mechanisms. Other reasons for concern about the possible consequences of ozone depletion have emerged. And there have been

many surprises! The overall aim of this Topic Study is to give you a feel for this unfolding story, and to highlight some of the intriguing and unresolved issues that remain. In doing so, the emphasis is largely, but not exclusively, on the science: the way an evolving scientific understanding of CFC-linked ozone depletion has influenced the policy-making process is also woven into the discussion.

STUDY COMMENT   The more chemical material in this Topic Study (mainly in Sections 3–5) draws on many of the general principles developed earlier in the Course, but the strongest links are with the kinetic analysis of reaction mechanisms discussed in Block 3. From this perspective, the story of stratospheric ozone depletion provides a good example of the way computer modelling studies are used to simulate the chemical behaviour of a complex natural system like the atmosphere – and some telling lessons about the uncertainties associated with such studies. However, there are also links with the material to come in Blocks 5 and 6, because research since the discovery of the ozone hole has highlighted the importance of chemical transformations on the surfaces of particles in the stratosphere. The video sequence *Ozone: the hole story* (band 4 on videocassette 1) provides an overview of the processes involved: it would be a good plan to watch it before you study Section 5.

Finally, two more general points before you get started. First, the material in this Topic Study is more descriptive, and less structured, than that in the main Blocks. Following the advice given in the 'study comments' should help you to engage with the text, and to focus on the most important points. Second, the 'boxed' material has a different role from that in the main Blocks. Here, we've used boxes to expand on points in the text, or to provide a little extra background information. With the notable exception of Box 1 (in Section 3.5), you may wish to treat these 'asides' as optional material.

2 THE ATMOSPHERE:
A BRIEF LOOK AT ITS STRUCTURE AND COMPOSITION

The ozone layer lies within the **stratosphere**, a region of the atmosphere that is strictly defined by reference to the rather complex temperature profile shown in Figure 1. The temperature falls with increasing altitude throughout the lower atmosphere or **troposphere**, reaching a minimum value of 200–220 K at the **tropopause**. This lies 8–15 km above the ground, depending (mainly) on latitude: specifically, it is higher, and colder, at the Equator than at the poles. Beyond the tropopause, the temperature starts to increase again, and continues to do so up to the stratopause (at about 50 km) – the upper 'boundary' of the stratosphere.

The names of the two lowest regions of the atmosphere reflect an important aspect of their physical behaviour – *tropos* is Greek for 'turning', and *stratos* is Latin for 'layered'. Put simply, the Earth's surface is warmed by absorbing solar radiation. Energy transferred from the surface to the overlying air warms the troposphere from *below* – whence the *negative* temperature gradient in this region. This situation sets the scene for the onset of convection, with warm surface air rising and cooler air aloft falling. As a result, the troposphere is characterized by strong vertical mixing.

By contrast, with warmer air lying above colder air, the stratosphere is inherently stable to convection. Indeed, it is sometimes said that the temperature inversion at the tropopause acts like a lid, separating the turbulent zone in which we live from the calm, stable stratosphere above. True, vertical transport up through the stratosphere is slow – with a time-scale of the order of years, rather than the days (at most) typical of the troposphere – and the mechanisms involved are more complex. In reality, however, the 'lid' is decidedly leaky. Thus, rapidly rising air can – and does – overshoot the tropopause, carrying its constituents into the stratosphere. Mostly this happens over the tropics, often in the updraught of violent storms. And there are return routes as well: tongues of stratospheric air descend into the upper troposphere at middle latitudes. In short, there is a constant, if relatively slow, exchange of air across the tropopause.

Away from the surface features that create turbulent flow close to ground level, there is also a recognizable large-scale pattern of horizontal motions in the lower atmosphere. This so-called 'general circulation' is the province of meteorology: it is manifest in the prevailing winds and weather systems experienced by different regions around the world. We shall not dwell on the details. Suffice it to say that the general circulation of the lower atmosphere is driven by the temperature contrast between low (warm) and high (cold) latitudes, but strongly influenced by the Earth's rotation about its axis. As a result, the regular wind regimes at all latitudes have a strong *zonal* (that is, east–west) component.

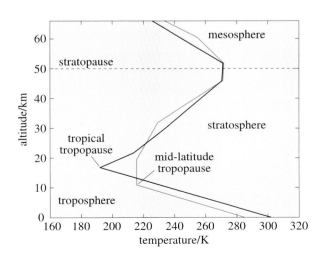

Figure 1 The vertical 'structure' of the atmosphere is defined by the way in which temperature varies with altitude. In each successive 'sphere' or zone, the temperature gradient is reversed. The profiles sketched here are typical for mid-latitudes in summer (the green line), and for tropical regions near the Equator (the black line).

For our purposes, the key point is that winds are responsible for transporting and mixing the chemical constituents in the air: they are the mechanism whereby substances released at the Earth's surface are distributed around the globe. From this perspective, it is important to appreciate that zonal mixing is generally much faster than is *meridional* (north–south) mixing. For example, mixing around a line of latitude typically takes about a month, whereas it takes about six months to achieve an even distribution throughout a hemisphere, and one or two years between the hemispheres. Finally, air moving up into the stratosphere joins the large-scale circulation that prevails in that region of the atmosphere. This has its own characteristic pattern of zonal and meridional flow – more on which later on.

There is one other important physical characteristic of the atmosphere that you need to be aware of – its pressure profile. The atmosphere gets progressively thinner or less dense with increasing altitude – that is, the *total* number (N) of molecules per unit volume (or **number density** = N/V) is lower, and so is the pressure (Figure 2). Because the fall-off is more or less exponential (notice the logarithmic scale on the pressure axis in the Figure), the bulk of the atmosphere is relatively close to the surface: 50% of the total mass is within some 5.5 km of the ground, and 99% lies below 30 km.

Assuming that air behaves as an ideal gas (i.e. that it obeys the equation $pV = nRT$), it should be clear that the total number density of air at a given altitude can be determined from temperature and pressure profiles like the ones in Figures 1 and 2. At ground level, where the globally averaged temperature and pressure are commonly taken to be $T = 288$ K and $p = 1\,013$ mbar, the value turns out to be:

$$[M] = 2.5 \times 10^{19} \text{ molecule cm}^{-3}$$

The designation [M] and the unit molecule cm^{-3} (or, strictly, just cm^{-3}) are those commonly employed by atmospheric scientists – as indeed is the use of millibars for pressure (see the caption to Figure 2): SAQ 1 (at the end of this Section) should help you to get used to working with these units.

Information on the chemical composition of the Earth's atmosphere is collected in Table 1. Here, the entry for each constituent records its **mixing ratio by volume**. For a given constituent (A, say), this is defined as:

$$\text{mixing ratio} = \frac{\text{number density of constituent}}{\text{total number density of air}} = \frac{N_A/V}{N/V} = \frac{N_A}{N} \tag{1}$$

Strictly, this definition produces a fraction: it represents the 'fractional abundance' of each species. For minor constituents, however, it is more convenient to quote mixing ratios as parts per million by volume (p.p.m.v.) or parts per billion (10^9) by volume (p.p.b.v.) – or parts per trillion (10^{12}) by volume (p.p.t.v.) for species present in truly trace amounts.

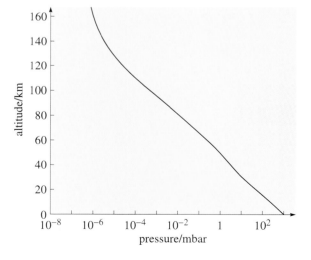

Figure 2   Pressure as a function of altitude in the Earth's atmosphere. Atmospheric scientists in general, and meteorologists in particular, have not yet adopted the SI unit of pressure (Pascal, Pa). Instead, the millibar (mbar) – familiar from weather forecasts – is almost universally used: 1 mbar = 10^2 Pa.

Table 1 A selection of the gases naturally present in the Earth's atmosphere, and the current (1990) average mixing ratio of each in the troposphere.

Gas	Mixing ratio
Major constituents	
nitrogen (N_2)	0.781 (78.1%)
oxygen (O_2)	0.209 (20.9%)
Minor constituents[a]	
water vapour(H_2O)	~0.01
carbon dioxide (CO_2)	353 p.p.m.v.
ozone (O_3)	0.01–0.1 p.p.m.v.
methane (CH_4)	1.72 p.p.m.v.
nitrous oxide (N_2O)	310 p.p.b.v.

[a] There are other minor constituents as well (notably argon, which accounts for nearly 1%), but the ones listed are those that play a part in the ozone story, both directly and *indirectly*: you will see later on that this is our reason for including CO_2.

The mixing ratios quoted in Table 1 are average values for the troposphere. Clearly, the *bulk* composition of this region is effectively a mixture of N_2 and O_2 in a 4 : 1 ratio. And this remains so throughout the stratosphere – and higher, up to altitudes of about 80 km. Beyond that, molecular diffusion begins to 'sort' the constituents on the basis of their molecular masses. But below 80 km, the bulk air motions outlined earlier are sufficiently rapid to ensure that the atmosphere is 'well-mixed' – or at least, it is as far as N_2 and O_2 are concerned. By contrast, the mixing ratios of minor constituents (or trace gases) can vary with all three spatial dimensions in the atmosphere – latitude, longitude and altitude – and sometimes with the seasons, or even with the time of day, as well. In particular, mixing ratios can show distinct, and sometimes very abrupt, changes at the tropopause. All of which brings us back to the ozone layer.

2.1   Ozone in the atmosphere

Ozone (O_3) occurs throughout the atmosphere, but only ever in trace amounts. Indeed, if all the ozone contained in the first 60 km or so of the atmosphere could be brought down and assembled at the Earth's surface, it would form a layer only some 3 mm thick. In practice, the bulk of the world's ozone (around 90% of it) is in the stratosphere. The profile sketched in Figure 3a is typical of the way in which the concentration (strictly, number density) of ozone varies with altitude. Notice that the highest concentrations occur in a well-defined layer at altitudes between about 20 and 35 km: this is the so-called **ozone layer**.

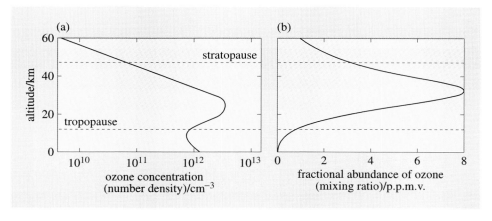

Figure 3   The vertical distribution of atmospheric ozone, expressed (a) as an absolute number density (concentration), and (b) as a relative mixing ratio (fractional abundance). The profiles sketched here would be typical for mid-latitudes.

■ As noted earlier, under average conditions at ground level, the *total* number density of air is around 10^{19} molecule cm^{-3}. Use Figure 3a to estimate the mixing ratio of ozone at ground level. Express your answer in p.p.m.v. and p.p.b.v.

▨ From the Figure, the ozone concentration at the surface is typically about 10^{12} molecule cm^{-3}, so the mixing ratio is $10^{12}/10^{19} = 10^{-7} = 0.1 \times 10^{-6}$ (0.1 p.p.m.v.) or 100×10^{-9} (100 p.p.b.v.).

Ground-level mixing ratios of ozone much higher than this are a serious problem, because it is a particularly unpleasant substance: even in very low concentrations, ozone irritates the respiratory system and can cause severe damage to human health. Plant growth may also be impaired. In industrial societies, one route by which ozone is generated close to ground level is by the action of sunlight on the mix of gaseous pollutants in vehicle exhausts. Ozone is one of the more noxious ingredients in 'photochemical smog', now recognized to be a global phenomenon that increasingly afflicts all major conurbations. These days, air quality is being taken seriously: in the UK, it is routinely commented on in weather forecasts.

In fact, ozone is produced and destroyed at a wide range of altitudes in the atmosphere, but the mechanisms involved are different in the different regions. Although we shall touch on the situation in the lower atmosphere from time to time, our central concern here is with the processes that control ozone concentrations in the stratosphere.

SAQ I   As a global and annual average, the temperature and pressure at an altitude of 25 km (the mid-stratosphere) can be taken to be $T = 219$ K and $p = 25.0$ mbar. Use the ideal gas equation to calculate:

(a) the total number density of air at 25 km;

(b) the concentration (number density) of O_2 at 25 km;

(c) the partial pressure of O_2 at 25 km.

Express your answers to parts (a) and (b) in the unit (molecule) cm^{-3}, and that to part (c) in both mbar and Pa.

3 THE STRATOSPHERIC OZONE CYCLE

The beneficial effects of ozone depend on the fact that most of it *is* in the stratosphere, well removed from direct contact with life. Figure 4a compares the spectral distribution of solar radiation reaching the Earth's surface with that outside the atmosphere. The point to register is the marked shift in the cut-off at the *short-wavelength* (ultraviolet, or uv) end of the spectrum: at ground level this lies at about 300 nm, whereas the distribution outside the atmosphere extends to wavelengths less than 200 nm. This shift is of critical importance to living systems. Radiation in the band labelled UV-C (100–280 nm) is lethal to microorganisms (whence its use in germicidal lamps), and energetic enough to damage proteins and other biological macromolecules, such as DNA. Up to about 240 nm, absorption by atmospheric O_2 (Figure 5) – in and above the stratosphere – provides an effective filter. But in the range 240–280 nm, protection from UV-C is due entirely to the ozone layer: O_3 happens to have an unusually strong absorption band at these critical wavelengths (Figure 5), and so such radiation cannot penetrate significantly below the mid-stratosphere. Weaker absorption by ozone in the band labelled UV-B (280–320 nm) attenuates the solar input at these wavelengths, but the effect is less complete: a fraction of UV-B penetrates all the way to the ground.

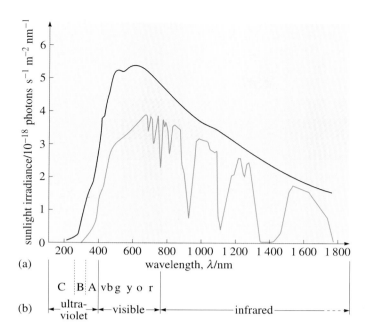

(a)

(b)

Figure 4 (a) Sunlight spectral irradiance (number of photons per second per square metre over each 1 nm wavelength interval). Black curve: distribution of solar irradiance outside the Earth's atmosphere. Green curve: distribution at the Earth's surface. The general reduction in intensity arises because some solar radiation is scattered directly back to space. The marked attenuation of *certain* wavelengths is due to absorption by chemical species in the atmosphere. (b) How part (a) relates to regions of the electromagnetic spectrum. By convention, the ultraviolet is split into three regions, A, B and C; colours in the visible region are indicated by initials.

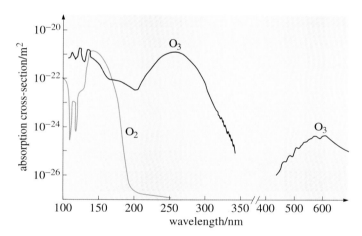

Figure 5 The variation with wavelength of the absorption of radiation by O_2 (green line) and O_3 (black line). The absorption cross-section is the molecular equivalent of the more familiar molar absorption coefficient (Block 2, Section 7.1.2).

Absorption of a photon at uv – or somewhat longer, visible – wavelengths leads to *electronic excitation*: for a molecule (AB, say) this can be represented as

$$AB + h\nu \longrightarrow AB^* \tag{2}$$

where $h\nu$ is the energy of the absorbed photon ($E = h\nu = hc/\lambda$). Electronic transitions in molecules are governed by the ordinary rules of spectroscopy. For absorption to occur, the fundamental requirement is that there exists an upper electronic state of the molecule that is separated from the ground state by an energy equal to that of the incident photon. Beyond that, there are certain selection rules, just as there are in vibrational and rotational spectroscopy: we shall not pursue the details. The important new feature is that electronic excitation can lead to *dissociation*, which amounts to the rupture of a chemical bond. This may come about when the energy of the absorbed photon exceeds the bond dissociation energy of the bond in question, as shown schematically in Figure 6. The **photodissociation** (or **photolysis**) of atmospheric constituents has one, all-important, *chemical* consequence: it generates highly reactive atomic and molecular fragments.

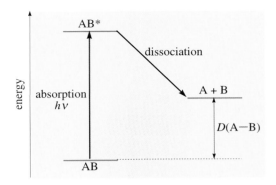

Figure 6 A schematic representation of the photochemical dissociation of a molecule AB, initiated by the absorption of a photon of energy hv. As shown here, excitation to an upper electronic state (AB*) with an energy greater than that necessary for bond dissociation, $D(A–B)$, leads to fragmentation. The excess photon energy is dissipated as kinetic energy of the fragment species, A and B.

3.1 A simple photochemical model: 'oxygen-only' chemistry

It is the photolysis of O_2 that initiates the production of ozone in the stratosphere. The process can be represented as follows:

$$O_2 + hv \longrightarrow O + O \quad (\lambda \leqslant 240 \text{ nm}) \tag{3}$$

which can be interpreted as implying that absorption of a photon with $\lambda \leqslant 240$ nm provides enough energy to break the O=O bond in O_2. In fact, absorption by O_2 at $\lambda < 175$ nm (Figure 5) generates an atomic fragment that is itself in an electronically excited state: we shall write it O*. Reactions involving O* come into our story later on. However, their almost exclusive fate is 'collisional deactivation', which returns them to the ground state. In the present context, no distinction need be drawn between O and O*.

Thus, we shall stick with equation 3, and simply take this to mean that the incoming solar radiation absorbed by O_2 provides a source of free oxygen atoms in (and above) the stratosphere. Once formed, oxygen atoms can combine with intact O_2 molecules to form ozone:

$$O + O_2 \longrightarrow O_3 \quad \Delta H_m^\ominus = -106.5 \text{ kJ mol}^{-1} \tag{4}$$

But ozone itself absorbs incoming solar radiation – not only in the vital uv range noted earlier, but also (albeit more weakly) at longer wavelengths stretching well into the visible band (Figure 5). For ozone, photons with wavelengths as long as 900 nm (where absorption begins) are energetic enough to cause dissociation. Once again, we represent the overall process as follows:

$$O_3 + hv \longrightarrow O + O_2 \quad (\lambda \leqslant 900 \text{ nm}) \tag{5}$$

but note, in passing, that absorption at $\lambda \leqslant 310$ nm yields excited O* atoms.

It is, then, apparent that the protection afforded by stratospheric ozone is sacrificial in nature. Concentrations build up if there is enough molecular *and atomic* oxygen around to provide numerous opportunities for the interaction in equation 4 to take place. In effect, ozone is pushed through the cycle of reactions captured in the centre of Figure 7: it absorbs radiation, breaking into its constituent parts, only to be formed again. This cycle is 'terminated' by reaction with a free oxygen atom:

$$O_3 + O \longrightarrow 2O_2 \quad \Delta H_m^\ominus = -391.9 \text{ kJ mol}^{-1} \tag{6}$$

Set in the context of your study of Block 3, the elementary steps in equations 3 to 6 should be viewed as one *plausible* candidate for the mechanism that is responsible for maintaining the ozone layer. This simple 'oxygen-only' scheme was first proposed as such in 1930, by the British geophysicist Sydney Chapman. It is now abundantly clear that matters are a good deal more complicated than this! Nevertheless, the so-called **Chapman mechanism** does capture the essence of the ozone budget in the stratosphere. There are two points to note.

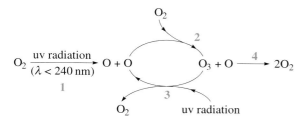

Figure 7 Ozone is created when short wavelength uv radiation breaks up an O_2 molecule (1), freeing its atoms to combine with other O_2 molecules (2). The ozone so formed is repeatedly broken up (3) and reformed (2), until it is destroyed by reaction with an oxygen atom (4).

First, if you think of atomic oxygen as the chain carrier, then the Chapman scheme contains the essential ingredients of a chain reaction. Specifically, there is an initiation step (reaction 3), which produces the chain carriers. The latter are recycled many times via reactions 4 and 5 – the propagation steps – and removed in a termination step (reaction 6).

■ In Block 3, we said that the propagation steps in a chain reaction cause the conversion of reactants into products: adding these steps together gives the overall stoichiometric equation. What do you get in this case?

▨ Nothing – i.e. adding equations 4 and 5 gives

$$O + O_2 + O_3 \longrightarrow O + O_2 + O_3$$

This somewhat curious result highlights an important difference between the Chapman scheme and the examples of chain reaction mechanisms that you met in Block 3. There, the underlying aim was to explain the process whereby reactants are converted into products. By contrast, the Chapman mechanism recognizes that ozone is constantly being created *and* destroyed in the stratosphere. The aim is to explain the dynamic balance between creation and destruction whereby the ozone concentration at a given altitude is maintained. The analysis of this **dynamic steady state** (sometimes termed a 'photostationary state') is taken up in the next Section.

Second, notice that the two purely chemical steps in the mechanism (reactions 4 and 6) are both strongly exothermic. As ozone cycles through its round of creation and destruction, the incoming solar energy absorbed by both O_2 and O_3 is ultimately released as heat. It is this *in situ* heating that is responsible for the temperature inversion at the tropopause – and hence for the very existence of the stratosphere. *In short, without the ozone layer there would be no stratosphere.*

3.2 A kinetic analysis of the Chapman mechanism

The shortcomings of the Chapman scheme are best exposed by comparing its predictions with the observed ozone profile (Figure 3). To do that, it is necessary to elaborate the mechanism into a *quantitative model*, such that the steady-state ozone concentration at different altitudes can be computed. And that requires information about the rate constants of the individual steps, collected as steps 1–4 below:

$$O_2 + hv \xrightarrow{\ j_1\ } 2O \qquad\qquad\qquad\qquad\qquad \textit{step 1}$$

$$O + O_2 + M \xrightarrow{\ k_2\ } O_3 + M \quad k_2 = 6 \times 10^{-34} \left(\frac{300\ \text{K}}{T} \right)^2 \text{cm}^6\,\text{s}^{-1} \qquad \textit{step 2}$$

$$O_3 + hv \xrightarrow{\ j_3\ } O + O_2 \qquad\qquad\qquad\qquad\qquad \textit{step 3}$$

$$O + O_3 \xrightarrow{\ k_4\ } 2O_2 \quad k_4 = 1 \times 10^{-11} \exp\left(-2\,100\ \text{K}/T\right) \text{cm}^3\,\text{s}^{-1} \qquad \textit{step 4}$$

As you know, information about rate constants comes from laboratory measurements, but there are several novel features here. First, the two photochemical steps are first-order processes, each characterized by a **photodissociation coefficient**, j (with the

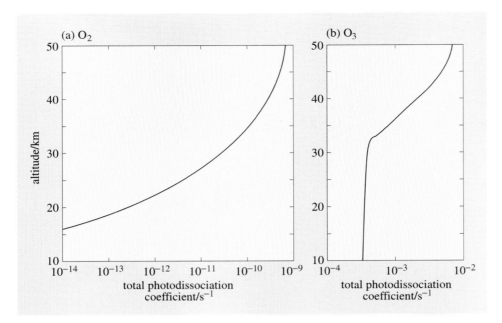

Figure 8 The total photodissociation coefficient as a function of altitude of (a) molecular oxygen, j_1; and (b) ozone, j_3. Notice the marked fall-off in j_1 with *decreasing* altitude, reflecting the progressive filtering out of photons energetic enough to photolyse O_2. By contrast, j_3 is less dependent on altitude: photolysis at the longer uv and visible wavelengths absorbed by O_3 persists throughout the stratosphere (and below). Higher values of j_3 at the top of the stratosphere are due to the strong absorption of wavelengths in the range 240–280 nm.

unit s^{-1}),* which is analogous to the rate constant k_R of an 'ordinary' (i.e. thermal) reaction. In the present context, values of j incorporate spectroscopic data (i.e. measurements of the wavelength-dependent absorption cross-section, as in Figure 5), together with information about the intensity distribution of incoming solar radiation and the way that this is attenuated as it passes down through the atmosphere. Plots of j_1 and j_3 as a function of altitude are shown in Figure 8.

The expressions given for k_2 and k_4 summarize the experimentally determined temperature-dependence of the rate constants for the thermal steps in the mechanism. Notice that we have now included a 'third body', M, in the first of these (step 2). Notice too that the rate constant for this *termolecular* step has an unfamiliar, *non-Arrhenius*, type of temperature-dependence. Indeed, the expression quoted indicates that the value of k_2 falls (slightly) with increasing temperature. Like atom recombination reactions (Block 2), those involving the combination of an atom and a small molecule – or indeed, two small molecules – invariably require collision with a third body to carry away the 'excess' energy of bond formation: otherwise the reactants would simply bounce apart again. Under these circumstances, high kinetic energies (increasing temperature) can be counterproductive.† In the context of atmospheric chemistry, M is invariably one of the most abundant molecules (N_2 or O_2), and the concentration of M is simply equated with the number density of air at the altitude in question (as discussed in Section 2).

Coming now to the analysis of the Chapman mechanism, your instinct is probably to embark on a steady-state treatment. For example, you might decide to treat the oxygen atom as a reactive intermediate, write an expression for $d[O]/dt$, set it equal to zero… and so on. SAQ 3 (at the end of this Section) invites you to do just that. Here, we adopt a different strategy – one that serves to illustrate a general device for simplifying complex kinetic schemes that is much used in modelling atmospheric chemistry.

---

* Try not to confuse the use of a *lower case j*, for photodissociation coefficient, with that of a *capital J*, for the rate of a reaction. Thus, for example, the rate of step 1 is given by $J_1 = j_1[O_2]$.

† More formally, theory suggests that the temperature-dependence of k for such termolecular reactions is better expressed in terms of a power, T^{-n}, rather than as a conventional activation energy. Expressions like the one quoted here for k_2 are obtained by fitting the experimental data against a T^{-n} law.

The starting point is the recognition that the individual steps in the mechanism operate on very different time-scales. Thus, throughout the stratosphere the photolysis of O_2 is intrinsically much slower than the photolysis of O_3: compare the values of j_1 and j_3 in Figure 8. Furthermore, it turns out that steps 2 and 3 result in a rapid interconversion of O and O_3 under stratospheric conditions. This rapid interconversion – implicit in the cycle we drew in Figure 7 – provides the rationale for introducing the concept of **odd oxygen**, O_x (O and O_3), as distinct from O_2: that is, 'odd' means 'an odd number of oxygen atoms'. To analyse the mechanism, it is useful to form the *sum* of odd oxygen – that is, $[O_x]$, the concentration of O_x, is defined as the sum $[O] + [O_3]$.

■   Use the Chapman scheme to write expressions for $d[O]/dt$ and $d[O_3]/dt$.

$$\frac{d[O]}{dt} = 2j_1[O_2] - k_2[O][O_2][M] + j_3[O_3] - k_4[O][O_3] \tag{7}$$

$$\frac{d[O_3]}{dt} = k_2[O][O_2][M] - j_3[O_3] - k_4[O][O_3] \tag{8}$$

Adding these expressions together gives:

$$\frac{d[O_x]}{dt} = \frac{d[O]}{dt} + \frac{d[O_3]}{dt} = 2j_1[O_2] - 2k_4[O][O_3] \tag{9}$$

This expression indicates that the rate of change of the concentration of odd oxygen is determined *solely* by the two slow steps in the mechanism – steps 1 and 4. That conclusion could have been arrived at by simply inspecting the information collected below: two odd-oxygen species (2O) are formed in step 1, and two (O and O_3) are lost in step 4. Steps 2 and 3 'do nothing' as far as odd oxygen is concerned: they interconvert O and O_3, but do not affect the *sum* of their concentrations.

		change in odd oxygen	
$O_2 + h\nu \xrightarrow{\ j_1\ } 2O$	*slow*	+2	*step 1*
$O + O_2 + M \xrightarrow{\ k_2\ } O_3 + M$	*fast*	0	*step 2*
$O_3 + h\nu \xrightarrow{\ j_3\ } O + O_2$	*fast*	0	*step 3*
$O + O_3 \xrightarrow{\ k_4\ } 2O_2$	*slow*	−2	*step 4*

Remember that the aim is to calculate the steady-state ozone concentration at a given altitude. To do that, two assumptions are made.

1   It is assumed that steps 2 and 3 effectively comprise a rapidly established (photochemical) equilibrium, such that

rate of step 2 = rate of step 3

or

$$k_2[O][O_2][M] = j_3[O_3] \tag{10}$$

so

$$[O]/[O_3] = j_3/k_2[O_2][M] \tag{11}$$

This assumption should be interpreted to mean that the individual concentrations of O and O_3 adjust rapidly to the much slower changes in the sum of their concentrations (given by the expression in equation 9). In other words, steps 2 and 3 act to *partition* the odd-oxygen species, and hence determine the ratio $[O]/[O_3]$ at a given altitude (equation 11).

2   It is assumed that the *overall* steady state at each altitude is defined by the condition $d[O_x]/dt = 0$. According to equation 9, this implies

$2j_1[O_2] = 2k_4[O][O_3]$ *at steady state*

Substituting for [O] from equation 11 ([O] = $j_3[O_3]/k_2[O_2][M]$) and rearranging, gives

$$j_1k_2[O_2]^2[M] = j_3k_4[O_3]^2$$

so

$$[O_3] = [O_2](j_1k_2[M]/j_3k_4)^{1/2} \qquad (12)$$

or, since [O_2] = 0.2[M] (SAQ 1),

$$[O_3] = 0.2[M]^{3/2}(j_1k_2/j_3k_4)^{1/2} \qquad (13)$$

Of the rate coefficients collected in the parentheses, it turns out that the temperature-dependence of k_2 and k_4 – together with variations in j_3 (Figure 8) – has only a relatively small effect on the *shape* of the calculated ozone profile: the dominant influence comes from the altitude-dependence of j_1 and [M]. Because these change with altitude in opposite ways, the computed ozone distribution mimics the characteristic bulge in the observed profile surprisingly well (Figure 9) – *except that it is too large*.

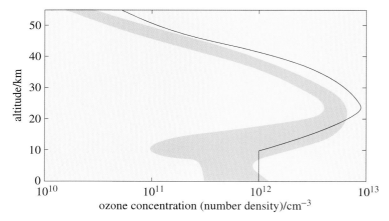

Figure 9   The concentration (number density) of ozone as a function of altitude. The shaded band represents the range of observed values. The black curve is the steady-state profile computed according to the Chapman 'oxygen-only' scheme (i.e. equation 13).

One final point is worth noting at this stage. According to the observational data recorded in Figure 10, O_3 is the dominant form of odd oxygen over the altitude range sampled in these measurements (the mid-stratosphere). This trend is even more apparent lower down in the stratosphere: at 15 km, for example, values of [O] are typically around 10^5 cm^{-3}, whereas [O_3] is about 10^{12} cm^{-3} (recall Figure 3).

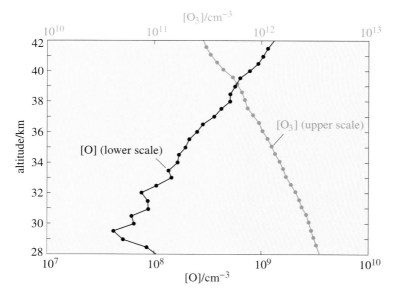

Figure 10   Concentrations of O and O_3 measured simultaneously within the same element of the stratosphere. Notice the very different concentration scales for the two species.

SAQ 2 According to the data in Figure 10, how does the ratio $[O]/[O_3]$ vary with altitude in the stratosphere? In general terms, is this variation consistent with the partitioning of the odd-oxygen species via the expression in equation 11?

STUDY COMMENT The following SAQ invites you to compare the analysis of the Chapman mechanism in this Section with a treatment that does not involve explicit use of the concept of odd oxygen. Make sure you try it at some stage.

SAQ 3 (a) According to the steady-state approximation (Block 3), if atomic oxygen behaves as a reactive intermediate, then it is legitimate to write $d[O]/dt = 0$. Further, the steady state for ozone can be defined by the condition $d[O_3]/dt = 0$. Confirm that, taken together, these two conditions again imply that

$$k_4[O][O_3] = j_1[O_2]$$

(b) You now have the problem of coming up with an expression relating $[O]$ and $[O_3]$, which is the point at which *this* analysis gets a bit tricky! As a first step, confirm that the equality above allows the expression for $d[O_3]/dt$ to be rewritten as:

$$\frac{d[O_3]}{dt} = k_2[O][O_2][M] - j_3[O_3] - j_1[O_2] = 0$$

Compare this expression with the one we obtained by treating steps 2 and 3 in the mechanism as a rapidly established equilibrium (equation 10). Can you now spot the implicit assumption behind that treatment?

3.3 The influence of trace species: catalytic cycles

The discrepancy evident in Figure 9 was recognized by the mid-1960s. With improved measurements, both in the laboratory and in the atmosphere, it became apparent that the Chapman scheme seriously underestimates the rate of *loss* of odd oxygen: step 4 in the mechanism removes only about 25% of the odd oxygen produced by oxygen photolysis (step 1). To keep stratospheric ozone concentrations at the observed levels, an additional – *and faster* – mechanism for destroying odd oxygen is required. A clue to the type of process involved came with the discovery that laboratory measurements of the rate of step 4 could be affected by the presence of trace impurities reacting in a cyclic manner.

The central idea is that trace species in the stratosphere – present at levels of a few p.p.b.v., at most – can engage in chain reactions that have the overall effect of *catalysing* step 4. Put simply, this can be represented as follows:

$$X + O_3 \longrightarrow XO + O_2 \tag{14}$$
$$XO + O \longrightarrow X + O_2 \tag{15}$$

net: $\quad O + O_3 \longrightarrow 2O_2$ *step 4*

with the 'catalytic' species X emerging unscathed at the end of the cycle – free to destroy more odd oxygen. The pair of reactions in equations 14 and 15 is commonly referred to as a **catalytic cycle**: using the notation introduced in Block 4, it can be represented as shown in Figure 11.

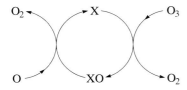

Figure 11 A representation of the catalytic cycle whereby X catalyses the loss of odd oxygen ($O + O_3 \longrightarrow 2O_2$).

Research since the 1960s has identified several candidates for the catalytic X. The most important of these in the *natural* stratosphere are believed to be X = HO· (the hydroxyl radical), X = NO (nitric oxide), and X = Cl· (atomic chlorine). We'll return to the natural sources of these species in Section 3.5: for now, it is sufficient to note that each of them contains an unpaired electron – that is, they are all *radicals*, as are the corresponding oxygenated species, XO = HO$_2$· (the hydroperoxyl radical), XO = NO$_2$ (nitrogen dioxide), and XO = ClO· (chlorine monoxide). As a result, X and XO are all highly reactive, which is clearly an important general criterion if the *indirect* route (via reactions 14 and 15) is to prove an effective means of destroying odd oxygen.

Expressed in a more quantitative way, the analysis in the previous Section can be repeated for an 'extended' Chapman mechanism that includes reactions 14 and 15 as steps 5 and 6. This can be written as follows:

			change in odd oxygen	
Chapman scheme	$O_2 + hv \xrightarrow{j_1} 2O$	*slow*	+2	*step 1*
	$O + O_2 + M \xrightarrow{k_2} O_3 + M$	*fast*	0	*step 2*
	$O_3 + hv \xrightarrow{j_3} O + O_2$	*fast*	0	*step 3*
	$O + O_3 \xrightarrow{k_4} 2O_2$	*slow*	−2	*step 4*
catalytic cycle	$X + O_3 \xrightarrow{k_5} XO + O_2$	*fast*	−1	*step 5*
	$XO + O \xrightarrow{k_6} X + O_2$	*fast*	−1	*step 6*

It can then be shown (see SAQ 4) that *each* catalytic cycle contributes an *extra* loss term to the Chapman expression for d[O$_x$]/dt (equation 9), as:

$$\frac{d[O_x]}{dt} = 2j_1[O_2] - 2\{k_4[O][O_3] + k_6[XO][O]\} \tag{16}$$

$\underbrace{\text{Chapman scheme}}_{\text{production of } O_x} \quad \overbrace{\text{catalytic cycle}}^{\text{loss of } O_x}$

where k_6 is the rate constant for any one of the reactions represented by step 6 in our extended mechanism. According to this expression, the contribution made by each of the **catalytic 'families'** – commonly designated **HO$_x$**, **NO$_x$** and **ClO$_x$** – depends on two factors: the concentration of the species XO, and the corresponding value of k_6.

The influence of the second factor is apparent in the experimental rate parameters collected in Table 2. Here, the first row refers to the *direct* reaction between O and O$_3$ (step 4, that is): notice that the activation energy is much higher than that for any of the other reactions with atomic oxygen (XO + O, step 6). Hence the striking contrast between the values of k_R calculated at 220 K – a temperature typical of the mid-stratosphere (SAQ 1) – listed in the final column of the table: the rate constant for the direct reaction (k_4) is orders of magnitude smaller than the corresponding value of k_6 for any of the catalytic reactions. In short, it is clear that the latter *can* make a significant contribution to the loss of odd oxygen – in principle, at least. The extent to which they do *in practice* then depends on the first factor noted above: the relative concentrations of the various XO species in the stratosphere.

STUDY COMMENT   The following SAQ provides an opportunity for you to practise using the concept of odd oxygen to analyse an extended Chapman mechanism.

SAQ 4

(a) Use the extended Chapman mechanism in the text to write an expression for d[O$_x$]/dt.

(b) Now apply a steady-state treatment to the reactive species X (or to XO), and hence show that your expression for d[O$_x$]/dt is consistent with that given in equation 16.

Table 2 Rate parameters for reactions involved in destroying odd oxygen.[a]

X	XO	Reaction	$\dfrac{A}{cm^3\ s^{-1}}$	$\dfrac{E_a}{kJ\ mol^{-1}}$	$\dfrac{k_R\ (220\ K)}{cm^3\ s^{-1}}$
		$O + O_3 \longrightarrow 2O_2$	1.0×10^{-11}	17.5	7.2×10^{-16}
HO·	HO$_2$·	$HO_2· + O \longrightarrow HO· + O_2$	3.0×10^{-11}	~ −2	7.4×10^{-11}
NO	NO$_2$	$NO_2 + O \longrightarrow NO + O_2$	6.5×10^{-12}	~ −1	1.1×10^{-11}
Cl·	ClO·	$ClO· + O \longrightarrow Cl· + O_2$	3.0×10^{-11}	−0.6	4.1×10^{-11}

[a] In each case, the temperature-dependence of the rate constant k_R has been fitted to an Arrhenius-type expression, $k_R = Ae^{-E_a/RT}$. Negative values of the activation energy, E_a, simply imply that k_R falls (slightly) with increasing temperature. $k_R(220\ K)$ is the rate constant calculated at 220 K.

3.4 A closer look at the catalytic 'families'

The notion of a catalytic cycle is a compelling one: it provides insight into the way in which minute traces of certain atoms and radicals can have a marked effect on the ozone budget. But it is far from the whole story. For a start, the membership of each of the catalytic families is wider than that indicated in the previous Section. Further, the families are *not independent* of one another: rather, they are coupled together by reactions between the members of different families. Because of this, the rate of ozone destruction for one cycle of reactions often depends on the concentration of a catalytic species that strictly 'belongs' to another family.

To illustrate the point, consider Figure 12: gathered here are most (but not all!) of the members of the 'extended' nitrogen family. I don't plan to embark on an exhaustive survey of the information summarized in the Figure. Rather, the object is to highlight a few key features: they relate to the parts picked out in colour.

Notice first of all that there are several processes that act to interconvert the NO$_x$ species – not just the reactions implicated in our catalytic cycle. Specifically, photolysis of NO$_2$ yields NO; nitric oxide is converted back into NO$_2$ by reaction with radical species (HO$_2$· and ClO·) that belong to the hydrogen and chlorine families – one example of coupling within the overall HO$_x$–NO$_x$–ClO$_x$ system.

Of more general importance, however, are processes (again highlighted in colour in Figure 12) whereby active radicals become 'locked up' as more stable molecules. Take the formation of N$_2$O$_5$ for example. Sooner or later, N$_2$O$_5$ breaks down again, via photolysis or thermal decomposition (as indicated in Figure 12). But this is a relatively slow process in the stratosphere: in the meantime, N$_2$O$_5$ acts as an unreactive *reservoir* of the nitrogen oxides that are active in destroying ozone.

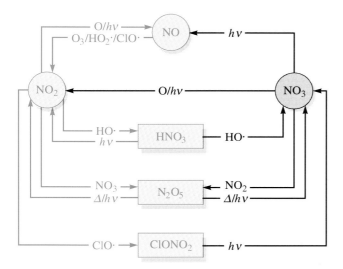

Figure 12 A schematic representation of the chemical 'relationships' among members of the nitrogen family. Here, active radical species are in circles; more stable reservoir molecules are in rectangles. The reaction partner that effects a transformation is included within the arrows; photochemical change is represented by *hv*; thermal decomposition by *Δ*.

Reservoir molecules formed by couplings *between* the families turn out to be even more important. Of the two such species included in Figure 12, nitric acid (HNO_3) acts as a major store of active nitrogen: on average, about half of the total stratospheric load of NO_x is held in this reservoir. Chlorine nitrate ($ClONO_2$) provides a second (albeit more temporary), holding tank for active nitrogen – *and it ties up active chlorine as well*. However, there is now good evidence that hydrogen chloride (HCl) acts as the main reservoir of active chlorine: indeed, some 70% of stratospheric chlorine is thought to be present as HCl. It is formed (and broken down again) via reactions that link together the ClO_x and HO_x families:

$$Cl\cdot + CH_4 \longrightarrow HCl + CH_3\cdot \qquad (17)$$

$$HCl + HO\cdot \longrightarrow Cl\cdot + H_2O \qquad (18)$$

In keeping with the notation adopted in Figure 12, this so-called 'holding cycle' can be represented as follows:

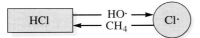

Storage systems of the type outlined above are of paramount importance in stratospheric chemistry: in effect, formation of the reservoir molecules diverts potentially catalytic species from 'active' to 'inactive' form. To a certain extent, this represents a permanent loss of active radicals. Transport of stratospheric air across the tropopause (Section 2) carries with it catalytic species locked up as reservoir molecules that are *soluble* (notably, HCl and HNO_3), and hence can be 'rained out' in the lower atmosphere. This provides a natural sink for the catalytic families. But in the main, the active radicals are eventually released again within the stratosphere, either chemically or photochemically (Figure 12). The crucial point is that the holding cycles exert a strong influence over the abundance (and vertical distribution) of the various catalytic species – and hence act to keep their ozone-destroying potential in check: see Figure 13. As you will see later, there is now good evidence that the release of active radicals – via *unexpected* routes – is the key to the special chemistry of polar ozone loss.

The brief outline above should have given you a feel for the complexity of stratospheric chemistry. But still we have barely scratched the surface. In reality, competitive processes abound, and there are countless other interactions among the atoms, radicals and molecules that control ozone concentrations in the stratosphere.

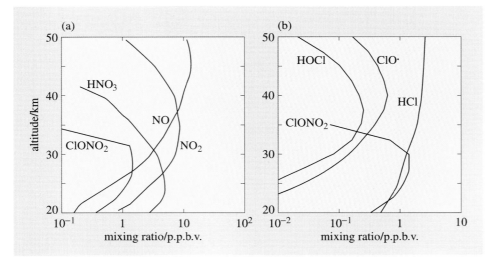

Figure 13 Vertical distribution of members of (a) the nitrogen family, and (b) the chlorine family, as measured at mid-latitudes at the end of April 1985. They are included here for illustrative purposes: imagine what the stratospheric load of active radicals might be like without the restraining influence of those holding cycles.

STUDY COMMENT   The following SAQ gives you a chance to collect together some of the important reactions that link the active chlorine radicals (Cl· and ClO·) and their reservoirs (HCl and ClONO$_2$) in the stratosphere. It would be a good plan to try it (and check our answer) before moving on.

SAQ 5   Figure 14 provides a framework for a schematic representation of ClO$_x$ chemistry, comparable with that for the nitrogen family in Figure 12. Drawing on information in this and previous Sections, identify the reaction partners (or other symbols) indicated by the letters a, b/c, d, e and f, and write an equation for each process.

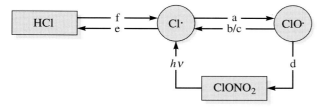

Figure 14   For use with SAQ 5.

3.5   Source gases – natural and unnatural

There have always been natural sources of the catalytic species discussed earlier: this is what keeps the natural rate of ozone production in check, and the ozone budget balanced at the 'normal' level. These **source gases** are again trace constituents in the atmosphere, produced – or emitted – at ground level, and then transported up into the stratosphere. They carry the elements of the catalytic families (nitrogen, hydrogen and chlorine), but in the form of relatively stable molecules that are usually rather insoluble in water, thus ensuring that they are not rapidly rained out in the troposphere. Information on the main *natural* source gas for each family is collected in Table 3: notice that they are all largely (or exclusively) of biological origin. Their fates once released to the atmosphere are reflected in the **atmospheric lifetimes** – *and here you should consult Box 1 (p. 24)* – recorded in the table.

Table 3   Tropospheric mixing ratios and lifetimes of natural source gases, together with the process that releases catalytic species in the stratosphere.

Source gas	Mixing ratio	Lifetime/yr	Release of active radicals via:
NITROGEN FAMILY			
nitrous oxide (N$_2$O)	~310 p.p.b.v.	~150	$N_2O + O^* \longrightarrow 2NO$

N$_2$O is part of the natural nitrogen cycle, released by microbial processes in soils and in the oceans.

HYDROGEN FAMILY			
methane (CH$_4$)	~1.72 p.p.m.v.	~10	$CH_4 + O^* \longrightarrow HO\cdot + CH_3\cdot$
			$CH_4 + HO\cdot \longrightarrow H_2O + CH_3\cdot$
			$\downarrow O^*$
			$2HO\cdot$

CH$_4$ is released during the breakdown of organic matter by bacteria under anaerobic (i.e. oxygen-free) conditions, principally in the intestines of ruminant animals and in waterlogged soils (bogs, swamps, tundra, etc.), whence methane's common name of 'marsh gas'.

CHLORINE FAMILY			
chloromethane (CH$_3$Cl)	~0.6 p.p.b.v.	~1.5	$CH_3Cl + HO\cdot \longrightarrow Cl\cdot + products$
			(multistep)

CH$_3$Cl is mainly generated during wood-rotting and natural forest fires: some arises from traditional slash-and-burn agriculture in the tropics.

For reasons that we shall return to shortly, the long lifetime of N_2O is characteristic of species that have chemical (or photochemical) sinks in the stratosphere, but *not* lower down in the atmosphere. Thus, N_2O is essentially inert in the troposphere: the process that converts it into NO (Table 3) becomes effective only at higher altitudes, where there is a more abundant supply of excited O^* atoms, generated mainly by the photolysis of O_3 and O_2 (as noted in Section 3.1).

By contrast, the shorter lifetimes of CH_4 and CH_3Cl reflect the fact that these species *do* have chemical sinks in the troposphere: the hydrogen atoms render both molecules liable to attack by HO· radicals in the lower atmosphere (see Box 1). Nevertheless, methane lingers in the troposphere for long enough to be moved around by the large-scale atmospheric motions mentioned in Section 2. As a result, CH_4 is distributed fairly evenly throughout the troposphere and, like N_2O, it is also transported upwards into the stratosphere. Here, it affects the ozone budget in a number of ways, but the most direct of these is its role as the main stratospheric source of active HO_x radicals (see Table 3).

Turning to the chlorine family, here the main *natural* source gas has a pretty short atmospheric lifetime. Because of this, rather little of the chloromethane that enters the atmosphere (only some 10%) actually reaches the stratosphere, where the chlorine atom is eventually stripped off – free to enter its cycle of ozone destruction (Table 3 again). This natural input is now dwarfed by that derived from the many other *synthetic* organochlorine compounds that are accumulating in the troposphere from strictly anthropogenic sources: this is where the **chlorofluorocarbons** (**CFCs**) enter the scene.

■ On current estimates, the total **chlorine loading** of the troposphere – that is, the mixing ratio of Cl atoms effectively carried in organic compounds – was around 3.6 p.p.b.v. in the early 1990s. What proportion of this was due to anthropogenic emissions?

▨ Taking CH_3Cl to be the only entirely natural source (Table 3), at least $(3.6 - 0.6)$ p.p.b.v. = 3.0 p.p.b.v. or $(3.0/3.6) \times 100 = 83\%$ was anthropogenic.

Pertinent information on the main contributors to this anthropogenic element is collected in Table 4: ignore the last three entries for now. Concentrating on the CFCs, note that these are all *fully* halogenated saturated hydrocarbons, but they contain varying proportions of the three elements, C, Cl and F. The immediately striking feature is the very long atmospheric lifetimes of these compounds. This stems from the very characteristics that made CFCs seem so ideal (Section 1): they are stable, non-toxic, and non-flammable – in short, completely inert in the lower atmosphere. Despite prodigious research efforts, no evidence has emerged for any significant tropospheric sink for CFCs (although tiny amounts are taken up by the oceans). On the contrary, the observational programmes set up to monitor these (and other) trace gases around the globe have recorded a steady build-up in the atmospheric burden of CFCs. Some typical data are included in Figure 15: they continue a rising trend that was first detected in the early 1970s.

Table 4 Tropospheric mixing ratios (1990) and overall atmospheric lifetimes for a selection of synthetic halocarbons.[a]

	Formula	Mixing ratio p.p.t.v.	Lifetime yr
CFC-11	$CFCl_3$	255	65
CFC-12	CF_2Cl_2	470	130
CFC-113	CCl_2FCClF_2	70	90
CFC-114	$CClF_2CClF_2$	5	200
CFC-115	$CClF_2CF_3$	4	400
carbon tetrachloride	CCl_4	107	50
methyl chloroform	CH_3CCl_3	160	6
HCFC-22	CHF_2Cl	103	15

[a] The number used to identify each CFC can be decoded, but we shall not pursue the details. The trivial names for tetrachloromethane (carbon tetrachloride) and 1,1,1-trichloroethane (methyl chloroform) are invariably used in the literature.

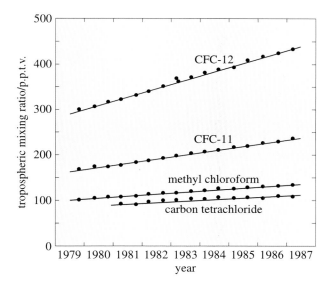

Figure 15 Trends in the tropospheric mixing ratios of several synthetic halocarbons, observed at mid-latitudes in the northern hemisphere.

The tropospheric inertness of the CFCs was the central thesis in the work published by Rowland and Molina in 1974. Given this, they argued that the only plausible fate for these compounds is transport upwards until they reach altitudes, *in and above the ozone layer*, where they eventually encounter the short uv wavelengths that are sufficiently energetic to break them up – in the range 190–210 nm for CFCs 11 and 12. Photolysis releases one Cl atom immediately, for example:

$$\text{CFC-11:} \quad CFCl_3 + h\nu \longrightarrow CFCl_2\cdot + Cl\cdot \tag{19}$$

$$\text{CFC-12:} \quad CF_2Cl_2 + h\nu \longrightarrow CF_2Cl\cdot + Cl\cdot \tag{20}$$

and the remaining Cl atoms as radical fragments engage in further chemical and photochemical reactions.

There are two points to note. First, only a small fraction of the total tropospheric mass is exchanged with the stratosphere each year. Or put slightly differently, it takes a long time for transport to cycle the tropospheric content of CFCs through to the appropriate altitude region for photolysis. And this produces a slow overall rate of removal from the atmosphere, whence the long lifetimes recorded in Table 4. Second, the operation of this high-altitude sink for CFCs (see Figure 16) adds to the natural background level of stratospheric chlorine, threatening to disturb the natural balance and cause a shift to a new steady state that sustains less ozone. For over ten years, assessments of the scale of that threat were based on the kind of modelling studies to which we now turn.

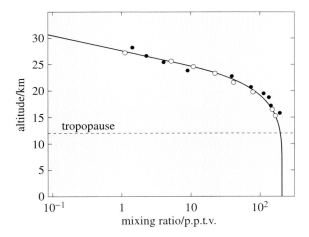

Figure 16 Evidence for the chemical decay of halocarbons like CFC-11 has been collected in experiments that compare their mixing ratios in the troposphere and the stratosphere. The data shown here were obtained at 44° N in October 1982 (white circles) and September 1983 (black circles). They strongly support the view that CFC-11 is removed from the atmosphere by processes that operate within the stratosphere – whence the marked fall-off with altitude in this region.

═══════════ **BOX 1** ═══════════

CHEMICAL TIME-SCALES AND ATMOSPHERIC LIFETIMES

1 Atmospheric lifetimes

Loss mechanisms or 'sinks' in the atmosphere – chemical or photochemical processes that destroy a particular species – operate on a variety of different time-scales. Atmospheric scientists characterize these time-scales by quoting the 'lifetime' of the species in question. In this context, it is *always* assumed that the loss mechanism follows first-order or, often, *pseudo*-first-order kinetics, that is:

A $\longrightarrow$ products

$$-\frac{d[A]}{dt} = k_R t$$

In Block 2, you met the integrated rate equation for a first-order reaction in logarithmic form:

$$\ln[A]_0 - \ln[A] = k_R t$$

Here, it is more illuminating to rewrite this expression in exponential form. This can be achieved by rearranging the expression above, as

$$\ln[A] - \ln[A]_0 = \ln([A]/[A]_0) = -k_R t$$

Taking 'inverse' natural logarithms, this becomes

$$[A]/[A]_0 = e^{-k_R t}$$

when it becomes apparent that *first-order kinetics implies that the concentration of A decays exponentially with time.*

The **lifetime** (symbol τ) of A is defined as the time it takes to reduce [A] to $1/e$ ($= e^{-1}$) of its initial value, $[A]_0$. Substituting $[A] = e^{-1}[A]_0$ into the expression above gives

$$e^{-1}[A]_0/[A]_0 = e^{-k_R \tau}$$

that is, $k_R \tau = 1$ or $\tau = 1/k_R$.

In practice, many (but not all) of the species released at the Earth's surface have chemical sinks in the lower atmosphere. Since the troposphere contains the bulk of the total atmospheric mass (see Section 2), chemical decay in this region effectively determines the overall atmospheric lifetime of such species.

2 An important tropospheric sink – attack by HO·

The HO· radical is the atmosphere's paramount molecular 'scavenger'. For example, CH_4 is oxidized in the atmosphere in a complex sequence of transformations that leads eventually to CO_2. But the whole process is triggered off by the following reaction:

$$CH_4 + HO· \xrightarrow{k_{CH_4}} CH_3· + H_2O$$

It is the rate of this hydrogen-abstraction reaction that effectively determines the atmospheric lifetime of CH_4. And the same goes for other hydrocarbons – and for their *partially* halogenated derivatives as well. Thus, for example, removal of chloromethane is initiated by:

$$CH_3Cl + HO· \xrightarrow{k_{CH_3Cl}} CH_2Cl· + H_2O$$

where k_{CH_3Cl} (like k_{CH_4}) is the corresponding bimolecular rate constant.

The atmospheric lifetimes (τ) of species subject to attack by HO· in the troposphere can be estimated as $\tau = 1/k_R'$, where k_R' is a *pseudo*-first-order rate constant, given in this context by $k_R' = k_R[HO·]$. For conditions typical of the troposphere – say $T = 250$ K (Figure 1) and $[HO·] = 10^6$ cm^{-3} – this produces the rough estimates for τ_{CH_4} and τ_{CH_3Cl} collected below (*cf* the values quoted in Table 3).

3 Sources of HO·

With abundant water vapour in the lower atmosphere (Table 1), the main *tropospheric* source of HO· is the following reaction:

$$O^* + H_2O \longrightarrow 2HO·$$

By contrast, *stratospheric* air is extremely dry – typical mixing ratios of H_2O are only a few p.p.m.v. Moisture-laden air *is* carried aloft in the lower atmosphere, but the temperature of the tropopause acts as an efficient 'cold-trap', effectively freeze-drying the air as it passes through to the stratosphere. In practice, it turns out that methane is the most important *source* of HO· – and of H_2O – in the stratosphere, as indicated in Table 3.

Species	Arrhenius expression for k_R	k_R at 250 K	τ/yr
CH_4	2.9×10^{-12} exp ($-1\,820$ K/T) cm^3 s^{-1}	2.0×10^{-15} cm^3 s^{-1}	~16
CH_3Cl	2.1×10^{-12} exp ($-1\,150$ K/T) cm^3 s^{-1}	2.1×10^{-14} cm^3 s^{-1}	~1.5

SAQ 6   In Block 2, you saw that the half-life, $t_{\frac{1}{2}}$, for a first-order reaction is given by $t_{\frac{1}{2}} = \ln 2/k_R = 0.693/k_R$. Given the definition of the lifetime, τ, in Box 1, it then follows that:

$$t_{\frac{1}{2}} = 0.693\,\tau$$

On this basis, estimate how long it will take for the atmospheric concentrations of CFC-11 and CFC-12 to be reduced to (a) a half, and (b) a quarter of the levels they have reached when emissions cease.

SAQ 7   Two of the halocarbons included in Table 4 have much shorter atmospheric lifetimes than the other compounds. Suggest a plausible explanation for this difference.

3.6   Summary of Sections 2 and 3

1   Ozone concentrations in the atmosphere vary, peaking in the stratosphere at around 20–35 km. This ozone layer acts as a filter, protecting life on Earth from the shorter uv wavelengths in the incoming solar radiation.

2   Fragmentation (photolysis or photodissociation) of a chemical species following absorption of solar radiation (at uv or visible wavelengths) is one of the most important photochemical processes in atmospheric chemistry. The rate of this first-order process is characterized by a photodissociation coefficient, j (with the unit s^{-1}).

3   According to the Chapman oxygen-only scheme (Figure 7), the steady-state ozone concentration at a given altitude results from a dynamic balance between the creation (via O_2 photolysis) and destruction of odd oxygen, O_x (O and O_3). Rapid photochemical equilibrium between O and O_3 acts to partition the odd-oxygen species, with O_3 dominating throughout the stratosphere. Heat released during the ozone cycle is responsible for the reversal in the Earth's temperature profile at the tropopause, the 'boundary' between the troposphere and the stratosphere.

4   Although qualitatively correct, the Chapman scheme seriously overestimates ozone concentrations in the stratosphere. In practice, the loss of odd oxygen ($O + O_3 \longrightarrow 2O_2$) is catalysed by various radical species, via the following cycle:

$$X + O_3 \longrightarrow XO + O_2$$
$$XO + O \longrightarrow X + O_2$$

where X may be HO·, or NO, or Cl· (or atomic bromine – see Exercise 1, below).

5   Interactions within, *and more importantly between*, the catalytic families (HO_x, NO_x and ClO_x) produce relatively unreactive reservoir molecules (such as N_2O_5, HNO_3, HCl and $ClONO_2$) that act as holding tanks for active radicals, keeping their ozone-destroying potential in check. Downward transport of soluble reservoir molecules (e.g. HNO_3 and HCl) acts as a natural sink for catalytic species.

6   Catalytic species are generated within the stratosphere by the breakdown – either photochemically (following absorption of the shorter uv wavelengths that don't penetrate to the lower atmosphere) or chemically (often by reaction with excited O^* atoms) – of molecules that are transported up from the troposphere. The main source gases for the nitrogen family (N_2O) and hydrogen family (CH_4) have natural (biological) origins at the Earth's surface. The principal source gases for the chlorine family are now non-natural, comprising a variety of synthetic halocarbons, typified by the CFCs.

7   Chemical or photochemical 'sinks' in the atmosphere operate on a variety of different time-scales. These are characterized by quoting the lifetime, τ, of the species in question – defined as the time required to reduce the concentration to $1/e$ of its original value (Box 1). The lifetime is related to the half-life of a first-order process as $t_{\frac{1}{2}} = 0.693\,\tau$ (SAQ 6). In practice, values of τ vary over an enormous range – from seconds or minutes (for highly reactive species like radicals), to years or a decade or so (for molecules, e.g. CH_4 and partially halogenated hydrocarbons, subject to chemical attack by HO· radicals in the troposphere), to several decades or even centuries (for compounds, such as the CFCs and N_2O, for which the only effective removal mechanism requires slow transport into the stratosphere).

In conclusion, Figure 17 summarizes, in a highly schematic and much simplified way, *some* of the important features of the overall system that controls stratospheric ozone.

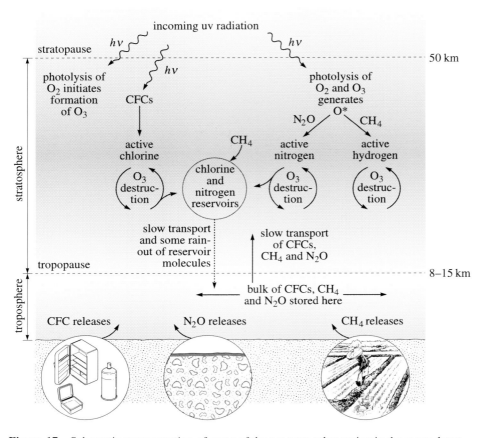

Figure 17   Schematic representation of some of the processes that maintain the ozone layer.

STUDY COMMENT   Do not miss out Exercise 1. It provides an opportunity for you to draw together and apply many of the ideas discussed in Section 3. It also revises your understanding of important general principles introduced in Blocks 1 and 2.

EXERCISE 1   The Montreal Protocol also seeks to control emissions of organic compounds containing bromine known as **halons**, widely used in specialist fire extinguishers. A typical example is CF_3Br (known commercially as halon-1301). Like other fully halogenated hydrocarbons, it is inert in the troposphere: photolysis in the stratosphere releases Br atoms.

(a)   Write a catalytic cycle showing destruction of odd oxygen by Br·. What information would you need in order to confirm that this is likely to be an effective loss mechanism? [*Hint* Refer back to the discussion in Section 3.3.]

(b)   Now concentrate on the thermodynamic and kinetic data collected in Table 5. Use inferences drawn from these data to discuss *briefly* the following proposition:

'Mole-for-mole, Br atoms released in the stratosphere represent a greater potential threat to the ozone layer than do Cl atoms'.

[*Hint* Start by commenting on the significance of information about the formation and breakdown of the species HX and $XONO_2$ in the stratosphere, and then use the data in Table 5 to compare the situation for X = Br with that for X = Cl.]

Table 5   Information on selected elementary reactions involving Cl and Br species, under stratospheric conditions.[a]

		Reaction	
X	$X + CH_4 \longrightarrow HX + CH_3$	$XO + NO_2 + M \longrightarrow XONO_2 + M$	$XONO_2 + h\nu \longrightarrow X + NO_3$
	$\Delta H_m^{\ominus}$ (298.15 K)/kJ mol^{-1}	k_R(220 K)/cm^6 s^{-1}	j/s^{-1} at 25 km
Cl	6.8	4.6×10^{-31}	7.4×10^{-5}
Br	72.5	1.3×10^{-30}	1.6×10^{-3}

[a] k_R(220 K) is the rate constant calculated at 220 K.

4 MODELLING STUDIES:
A HISTORICAL PERSPECTIVE

Figure 18 is a more complete (and somewhat less user-friendly!) representation of reactions within and between the three catalytic families that dominate stratospheric ozone chemistry. The *sole* reason for including it here is to convey something of the extraordinary complexity of this interconnected system – a network of coupled, and often competing, chemical and photochemical reactions. A full understanding of ozone chemistry involves the *simultaneous* consideration of the rates of all of these processes. Clearly, there is no simple way of doing that, and hence predicting ozone concentrations and the way these are likely to respond to enhanced levels of stratospheric chlorine – or indeed, to any other perturbation. The appropriate tool is a computer model. In this Section, we outline some of the assumptions and approximations that are commonly incorporated into such models. We do that in the context of a brief look at the evolution of model forecasts of ozone depletion, an essential backdrop for the material to come in later Sections. First, we need to know a little more about the influence of atmospheric motions.

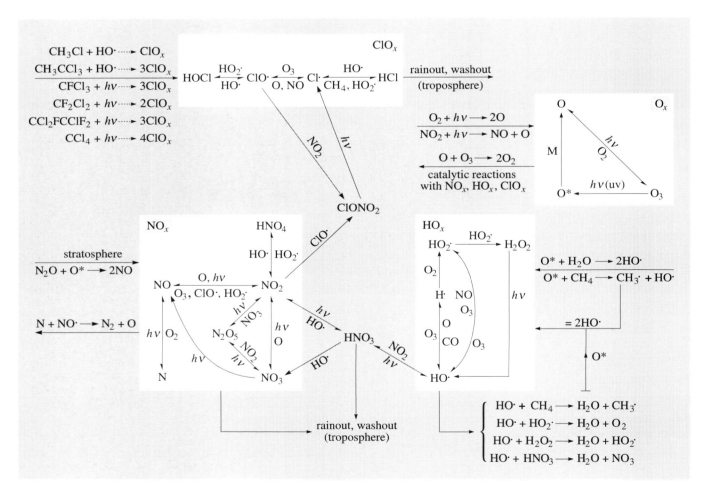

Figure 18   A fuller, but still incomplete, version of the stratospheric ozone chemistry captured schematically in Figure 17.

4.1 The global distribution of stratospheric ozone: the influence of transport processes

We have already touched on the importance of atmospheric motions to the ozone budget in the stratosphere. Source gases have to be carried into the stratosphere, and then move up through it. Relatively long-lived reservoir molecules (such as HCl and HNO$_3$) can be transported down into the lower atmosphere – and so can ozone itself: indeed, this is an important natural source of the 'normal' background level of ozone in the troposphere. But on top of this, the large-scale circulation of the stratosphere also plays a critical role in moving ozone around the world – and so determining its *global distribution*. There is evidence to this effect in Figure 19. Here, the 'contours' record measured values of the **total ozone column** (also known simply as **column ozone**) – that is, the total amount of ozone above a unit area of the Earth's surface. Values of column ozone contain no information about the vertical distribution of the gas: just the total number of molecules in a column stretching up from the surface to the top of the atmosphere. The data in Figure 19 are typical of years *before* the ozone hole started to appear, recorded as a function of both *latitude* (from the North Pole at the top to the South Pole at the bottom) and *season* (from January to December across the figure).

◼ Where and when do the highest amounts of ozone occur? Is this what you would have expected?

◻ According to Figure 19, column ozone is greatest at high latitudes in late winter and early spring – around the end of March in north polar regions (top left), and in mid-October at around 50° S (bottom right). This is surprising. Because the production of ozone is driven by sunlight, one might expect maximum amounts at low latitudes, where there is the greatest input of solar radiation: in practice, column ozone is lowest at equatorial latitudes.

In fact, ozone production *is* greatest above the Equator (at altitudes of some 30–40 km), but it is transported from there downward and toward the poles in a stratospheric circulation that is biased in favour of the 'winter' hemisphere, as suggested by the simplified picture in Figure 20.

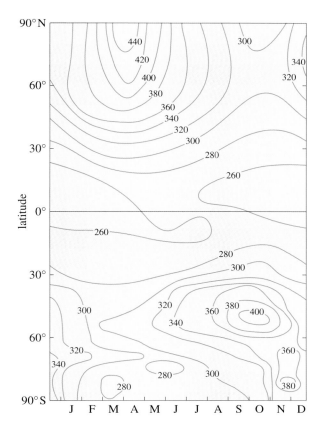

Figure 19 The global distribution of ozone, as a function of latitude and time of year. The numbers are average values of the total ozone column in Dobson Units (DU) averaged over all longitudes, based on data from a network of ground-based measuring stations around the world, for years prior to 1974. 100 DU represents the ozone in a column that would be 1 mm thick at the surface. The unit is named after Gordon Dobson, who developed the ozone spectrophotometer in 1930. Still widely used, the instrument measures the intensity of sunlight at two different wavelengths in the region 305–340 nm, where O$_3$ absorbs (Figure 5). One of the wavelengths is much less strongly absorbed than the other: from a knowledge of the absorption cross-section at the two wavelengths, together with the Beer–Lambert law (Block 2), the ozone column density can be deduced.

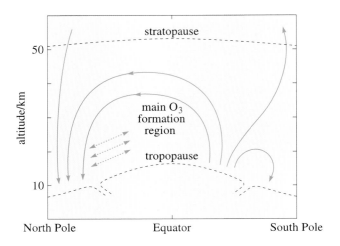

Figure 20 Ozone production and the circulation of the stratosphere, shown highly schematically for winter in the northern hemisphere. The flow is reversed during winter in the southern hemisphere – but the 'quasi-horizontal' transport (dashed green lines) is weaker and does not penetrate further south than about 60° S. The net effect of this circulation pattern is to move ozone poleward and downward from the tropical production region.

4.2 Formulating a model

The influence of atmospheric dynamics has important implications when it comes to the formulation of a model capable of simulating all the subtleties of ozone distribution – with altitude and latitude (and with longitude too, although this is less marked), and with the changing seasons. To illustrate the point, concentrate on Figure 21. This invites you to think of the atmosphere as being carved up into an array of boxes, defined by a grid spread over the surface of the globe. Now focus on what is going on inside one of these 'grid boxes' at some fixed location in the stratosphere, and then cast your mind back to the kinetic analysis of the simple Chapman scheme in Section 3.2. The implicit assumption behind the end result of that analysis (equation 13) was that the concentration of O_3 in this box is determined by the balance between the production and loss of odd oxygen. This is a valid assumption, *provided* that this overall photochemical steady state is established more quickly than the time-scale over which O_3 is transported into or out of the box (in any of three dimensions) by atmospheric motions. In practice, it turns out that this is a reasonable approximation above about 40 km, but an increasingly poor one lower in the stratosphere.

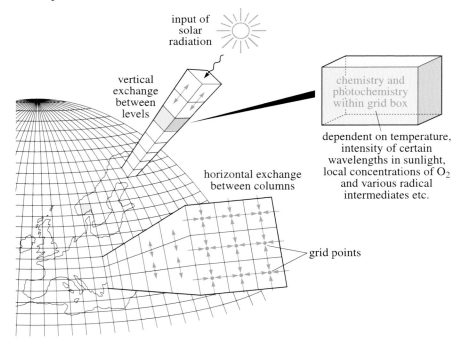

Figure 21 Schematic representation of the basic structure of a three-dimensional (3-D) atmospheric model that would be capable of simulating the variation of stratospheric ozone with altitude, latitude and longitude – and with the changing seasons.

The formulation of an 'ideal' model should, then, incorporate mathematical descriptions of *both* (photo)chemical *and* transport processes (operating in all three dimensions). In principle, all that is required is the writing of a time-dependent equation for the concentration of each chemical species *i* treated in the model (that is, an expression for $d[i]/dt$ at each grid point (Figure 21), taking into account transport processes, and recognizing that in such equations the production and loss rates will, in general, be complex functions involving photodissociation coefficients and rate constants (dependent, respectively, on solar irradiance and temperature) multiplied by the concentrations of the species involved (which effectively couples the equations together). Phew!

The resulting equations look, and are, extremely complex: solving (i.e. integrating) them, and hence computing the concentrations of all constituents – and how these vary in both space and time – is more complex still. Recall the number of chemical species and linked processes indicated in Figure 18. Even more daunting is the task implied by the phrase 'taking into account transport processes'. Basically, that amounts to simulating the large-scale circulation of the atmosphere and the ensuing transport of chemical constituents by winds and vertical mixing processes. That task is one of the principal goals of the theoretical meteorological community, through the design of 3-D weather prediction models and 'general circulation models' (GCMs). Its satisfactory achievement is only now coming within the modellers' grasp and, even then, only with dedicated use of the world's largest and fastest supercomputers.

At the time of writing (1995), the incorporation of detailed ozone chemistry into a GCM remains a largely unrealized goal, mainly because running such a model would be so demanding of computer time and memory. The problem is particularly pressing when a model is used to forecast the effect on the distribution of ozone of *changing* concentrations of, say, CFCs, because then demanding calculations must be repeated many times over. To keep the task down to manageable proportions, most such forecasts have used models that incorporate various simplifying assumptions and approximations. These fall into two broad categories:

- Techniques designed to reduce and simplify the set of rate equations to be solved – the procedure exemplified by forming the sum of odd oxygen ($[O_x] = [O] + [O_3]$) in Section 3.2. Often, two species (O and O_3 in this instance) undergo rapid interconversion, but much slower conversion into and/or from a third species (here O_2). Forming the sum of the two (i.e. writing an expression for $d[O_x]/dt$) causes the rapidly changing terms to cancel, leaving only reactions with longer time-scales (steps 1 and 4 in the Chapman scheme). Inclusion of the sum, rather than the individual components, then allows a more efficient numerical integration of the rate equations.

- Incorporating a much simplified treatment of transport by atmospheric motions, typically designed to reduce the number of spatial dimensions. More on this in the context of model forecasts.

4.3 Model predictions of global ozone depletion: changing perceptions

One-dimensional (1-D) models effectively treat the atmosphere as a single, globally averaged (over all latitudes and longitudes) column of air, divided into a series of layers in the vertical. There is thus no lateral transport of chemical constituents, and vertical transport at specific altitudes is represented in an average and approximate way. Given a particular temperature and pressure profile, input of solar radiation, etc., such models can simulate the vertical distribution of O_3 (and other species, like the catalytic radicals and reservoir molecules) – but not any lateral or seasonal variations. The altitude-profiles they produce are necessarily global averages.

The great strength of 1-D models is that they are computationally fast. As a result, they were widely used to forecast the effects on the total (average) ozone column of changing concentrations of the source gases discussed in Section 3.5. A typical study would involve running a *control simulation* with model inputs (concentrations of source gases and so on) appropriate to the then current conditions. The next step is to introduce the perturbation of interest (effectively a change in 'input' conditions, such as a specified increase in the chlorine loading of the atmosphere) and then run the model for many simulated years until the modelled atmosphere settles into a new steady state. The computed ozone column (and/or altitude profile) is then compared with that generated during the control run. Since 1974, 1-D model forecasts of ozone depletion by CFCs have fluctuated widely: Figure 22 presents a typical record through to the mid-1980s.

The fluctuations in Figure 22 reflect the many uncertainties associated with the modelling process, but many of the marked shifts recorded there can be largely attributed to changes in the chemistry 'built into' the model – specifically, to revised rate coefficients and to the inclusion of reactions that had been overlooked earlier. However, during the early 1980s, there was a trend to lower estimates of ozone depletion (evident in Figure 22), and this reflected the influence of another factor: a growing recognition that parallel changes in the atmospheric concentrations of certain other trace gases could act to *mitigate* the effect of CFCs on the total ozone column.

Observational programmes designed to monitor the composition of the atmosphere have revealed an ongoing accumulation not only of CFCs (and other chlorinated compounds), but also of various *natural* constituents. Chief among these are CH_4 and N_2O – two of the source gases identified in Section 3.5 – *and* CO_2. (The reason for including CO_2 in this list will become apparent in a moment.) The build-up of these gases is generally acknowledged to be an inadvertent by-product of human population growth and economic development: burning fossil fuels releases CO_2; rice paddies are major anthropogenic sources of CH_4; using nitrogenous fertilizers can enhance the flux of N_2O from soils; and so on. Suffice it to say that the 1980s saw 1-D models being used to explore the effect on column ozone of a range of hypothetical 'what if?' situations – or scenarios – involving projected future levels of these gases, as well as of the synthetic halocarbons.

Some typical results are collected in Figure 23. As noted in the caption, these forecasts come from the same modelling study (it was conducted in the mid-1980s), with the same 'unperturbed' model atmosphere: the precise details need not concern us. The crucial point is the *pattern* of change they predict. Thus, for example, a scenario that increases the total chlorine loading (to a large, but not unrealistic, level *in the absence* of controls over CFC emissions, see Section 7) does erode the ozone layer (Figure 23a) – as does a 20% increase in N_2O (Figure 23b). But notice the effect of doubling the concentration of CH_4 (Figure 23c). Because methane is the major source of radicals for the hydrogen family (Section 3.5), you might have again expected an overall loss of ozone. The striking feature is that this loss is restricted to the uppermost reaches of the stratosphere, and more than compensated by *increases* lower down. This reflects the complexity of methane's role in the atmosphere. In the lower stratosphere, increased CH_4 interferes with the effectiveness of the chlorine cycle: recall that it is instrumental in tying up active chlorine as HCl via reaction 17 (Section 3.4). Lower still, methane gets involved in just the same processes that *produce* ozone in photochemical smog (Section 2.1).

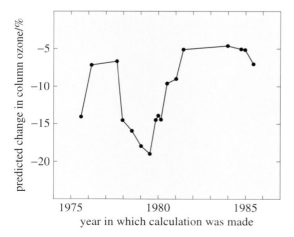

Figure 22 Record of calculations of the expected change in total column ozone (at steady state) from CFC release at the 1974 rate, as predicted by a particular 1-D model. The horizontal axis represents the year in which each calculation was made.

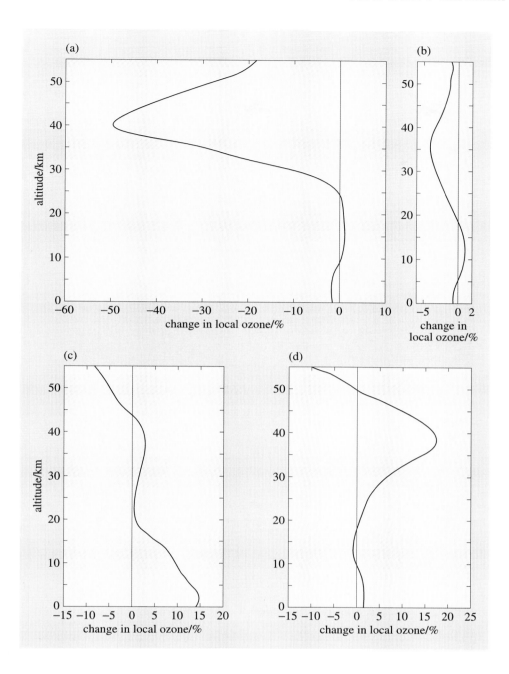

Figure 23 Altitude profiles of the percentage change in ozone (at steady state), predicted by a particular 1-D model, due to prescribed increases in the atmospheric concentrations of certain trace gases. In each case, the change is relative to a control run with an 'unperturbed' atmosphere containing a total chlorine loading of about 1.3 p.p.b.v. (a) Total chlorine increased to 8 p.p.b.v.; (b) N_2O increased by 20%; (c) CH_4 doubled; (d) CO_2 doubled. These levels of N_2O, CH_4 and CO_2 are expected to occur around the middle of the 21st century, under a 'business-as-usual' scenario: think of it as a continuation of past and current trends, assuming that no *active* steps are taken to curtail emissions of these gases.

Concentrate now on Figure 23d. CO_2 is not involved in *any* of the chemical cycles that control the ozone budget: indeed, CO_2 is chemically unreactive in both the troposphere and the stratosphere. So why should doubling its concentration have any effect on ozone? Put simply, it comes down to the fact that CO_2 is the archetypal 'greenhouse' gas. Enhanced levels of CO_2 in the atmosphere are expected to *cool* the stratosphere (Box 2), as well as leading to the better-known (if still much debated) warming of the troposphere. SAQ 8 (on p. 35) invites you to think through the link between a cooler stratosphere and the increased ozone concentrations evident in Figure 23d.

BOX 2

RADIATION 'TRAPPING' AND THE GREENHOUSE EFFECT

The overall temperature of the Earth is the result of a balance between the rate at which radiant energy comes in from the Sun and the rate at which it is radiated out again. If the Earth had no atmosphere, this balance would produce a global-mean surface temperature of around −18 °C. In practice, it is a comfortable 15 °C, because the atmosphere serves to retain heat near the surface. The key points are summarized in a highly schematic way in the Figure below.

In brief, the Earth intercepts solar radiation, with a spectral distribution that peaks in the visible (Figure 4). About 30% of this is reflected directly back to space, but the rest passes through the atmosphere without much absorption – penetrating to the surface, where it *is* absorbed. The warm surface, in turn, emits radiation, but with a spectral distribution that lies entirely at longer infrared wavelengths. Most of this radiation cannot pass *directly* out to space, because several atmospheric constituents absorb in the infrared. The most important of these are water vapour and CO_2, but the list also includes CH_4, N_2O and ozone. (The major atmospheric constituents, N_2 and O_2, are *not* infrared active, because they neither have a permanent dipole moment, nor do they acquire one when they vibrate.)

Some of the energy absorbed by these **greenhouse gases** is ultimately re-radiated out to space, thus maintaining a balance between the input and output of radiant energy at the top of the atmosphere. But some of the absorbed energy is retained in the lower atmosphere (vibrational energy being converted into kinetic energy through molecular collisions) or re-emitted back downward, thus increasing the energy input to the underlying surface. This phenomenon, which keeps the Earth's surface some 33 °C warmer than it would

otherwise be, is known as the **greenhouse effect**.

The ongoing increase in the atmospheric burden of natural greenhouse gases – together with the addition of *new* infrared-active species (notably the CFCs and other synthetic halocarbons) – is expected to lead to an enhanced radiative heating, and hence warming, of the surface and lower atmosphere. Whence the current concern about 'global warming', and other possible changes to the climate experienced by different regions around the world. Less well known is the expectation that this surface warming will be accompanied by a *cooling* of the stratosphere.

In reality, outgoing radiation of the pertinent wavelengths is repeatedly absorbed and re-emitted by different molecules of each gas as it 'works up' through the atmosphere. There is a net emission to space only from levels high enough for absorption to have become weak. Now have another look at the Figure. As indicated there, the existence of the ozone layer warms the stratosphere, not only because O_3 absorbs incoming uv (Section 3.1), but also because it absorbs certain wavelengths in the outgoing infrared radiation coming up from below. Roughly half of the absorbed energy is re-emitted – some out to space (from altitudes above 30 km), some back down to the troposphere (from the lower stratosphere): the remainder contributes to the *net* radiative heating of the stratosphere. This is balanced by the cooling effect of the *net* emission to space by other greenhouse gases in the stratosphere, the main contributor being CO_2. Additional CO_2 at these altitudes (an inevitable consequence of its build-up in the troposphere) is expected to enhance the net infrared emission to space, leading to a cooling, rather than warming, of the stratosphere.

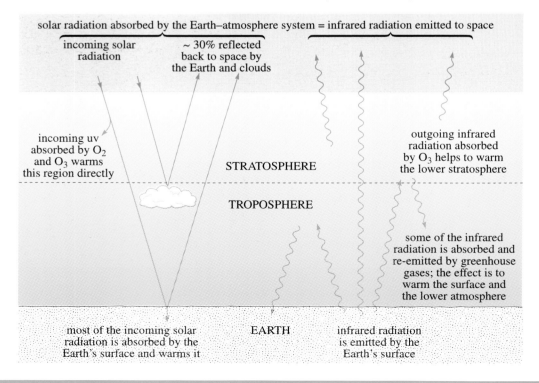

SAQ 8   To simplify matters, assume that ozone concentrations at a given altitude are controlled by the balance between the creation and loss of odd oxygen, according to a generalization of the expression you derived in SAQ 4, that is:

$$d[O_x]/dt = 2j_1[O_2] - 2\{k_4[O][O_3] + \Sigma k_6[O][XO]\}$$

where Σ represents a sum of terms, one for each of the catalytic cycles. Drawing on the information collected in Table 2 (Section 3.3), explain *briefly* why a cooling of the stratosphere might be expected to lead to enhanced ozone concentrations.

The discussion above serves to highlight two important points. First, in the real world, the concentrations of *all* the trace gases considered here are changing simultaneously. Modellers deal with that situation by conducting 'multiple perturbation' studies, in which the concentrations of the entire suite of important gases are varied at the same time. Secondly, *all* of the source gases identified in Section 3.5 are also greenhouse gases (Box 2), so any increases in their atmospheric abundance – as in that of CO_2 – will tend to alter the temperatures of the troposphere and stratosphere. And that affects the chemistry. Multiple perturbation studies using 1-D models that incorporate this 'temperature feedback' were a major factor in reducing forecasts of ozone depletion to more modest levels.

BOX 3

MODELLING AND MEASUREMENT

When a model is used to simulate conditions in the stratosphere, the results include not only the ozone data, but also information about the concentrations of the trace constituents that play a crucial part in the postulated chemical schemes as active radicals – and their reservoirs. The better the match with observational data on these species, the greater the confidence in the completeness of the chemical scheme, and in the credibility of the simplifying assumptions and approximations incorporated in the model. Many techniques now exist for the measurement of minor species in the atmosphere: often, they involve absorption and/or emission spectroscopy in spectral regions ranging from the uv to the microwave. Increasingly, ground-based

measurements have been supplemented by instruments carried on balloons, aircraft and rockets: satellites and space-shuttle platforms have been harnessed to provide global coverage.

For illustrative purposes only, the Figure below shows a comparison between model-simulated (green) and observed (black) mixing ratios for key species in the (a) nitrogen and (b) chlorine families. The comparison relates to conditions at 30° N at the end of April 1985, the measured values being those included in Figure 13. By the mid-1980s, comparisons like this had produced a measure of confidence that the general pattern of stratospheric ozone chemistry was quite well understood. Events were soon to prove otherwise.

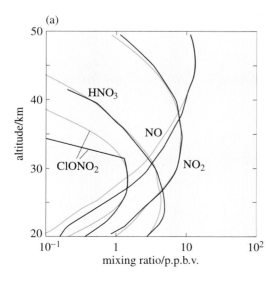

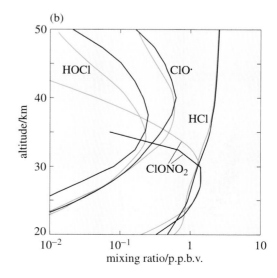

For all the valuable insights provided by 1-D models, they contain only a very crude description of atmospheric dynamics. Better in this respect are *two-dimensional (2-D)* models, which incorporate elements of transport in the meridional (north–south) direction, as well as in the vertical. The underlying rationale here is that zonal (east–west) mixing is much faster, so that chemical composition changes much less with longitude than it does with latitude and altitude (Section 2). In practice, 2-D models can simulate the general pattern of global ozone distribution shown in Figure 19 remarkably well. By the mid-1980s, this type of 'validation' study (see also Box 3) had produced a measure of confidence in *forecasts* made with 2-D models. There was an emerging consensus that continued release of CFCs at the then current levels (along with other source gases *and* CO_2) *would* erode the ozone layer – but the agreed figures at the time suggested a relatively slow attrition of something less than 1% of global ozone per decade. Within these global figures there was some indication that ozone loss would probably be greatest at high latitudes in winter. But nowhere was there any hint of a sudden and dramatic decline on anything like the scale that has actually occurred. In particular, none of the models predicted the ozone hole over Antarctica.

4.4   Summary of Section 4

To simulate all aspects of the variation of stratospheric ozone, spatially and with the seasons, strictly requires a sophisticated 3-D GCM that incorporates detailed ozone chemistry. Because of computational limitations, most assessments of the threat to ozone posed by human activities used models (1-D or 2-D) that contain a much simplified treatment of atmospheric dynamics. Through to the mid-1980s, such model predictions showed a trend to lower estimates of ozone depletion, partly because of a growing understanding that an ongoing accumulation of other trace gases in the atmosphere (notably CH_4 and CO_2) could act to mitigate the effect of CFCs on the total ozone column. All models failed to predict the Antarctic ozone hole.

SAQ 9   Refer back to the results of the modelling study that increases the total chlorine loading of the atmosphere shown in Figure 23a.

(a)   Under this scenario, ozone loss is predicted to be greatest at altitudes between about 35 and 45 km. Explain why this should be so. [*Hint* Think about the vertical distribution of chlorine species evident in Figure 13 and in the diagram in Box 3.]

(b)   According to Figure 23a, the peak ozone loss at 40 km could be as high as 50%, and yet the predicted change to the *total* ozone column in this study was only −5.7%. Try to rationalize these two results. [*Hint* Refer back to the ozone profile shown in Figure 3a.]

5 NEW SCIENCE: NEW URGENCY

STUDY COMMENT   This would be a good point to view the video sequence associated with this Topic Study (band 4 on videocassette 1). As well as providing an overview of the chemistry involved in polar ozone loss, it also includes computer-generated images of the meteorological phenomenon (known as the polar vortex) that is central to the discussion in Section 5.1.2.

Large losses of total ozone in Antarctica reveal seasonal ClO_x/NO_x interaction. (J.C. Farman, B.G. Gardiner and J.D. Shanklin, *Nature*, 16 May 1985)

The measurements from the British Antarctic Survey (BAS) station at Halley Bay reported by Joe Farman and his colleagues in 1985 are included in Figure 24. There is no doubt that this report marked a crucial turning-point in the debate about the threat posed by continued release of CFCs. Up to this point, that debate had turned, almost exclusively, on the results of modelling studies, *not* actual observations. True, experimental programmes had provided strong evidence for the breakdown of CFCs in the stratosphere (recall Figure 16), and there was reasonably good agreement between the available observational data and the calculated concentrations of the catalytic species, and their reservoirs (see Box 3, for example). But there remained a vital area of uncertainty: was anything actually happening to the ozone layer?

Here, the underlying problem is that column ozone is naturally highly variable. Within the tropics, the average figures recorded in Figure 19 present a typical picture, with small but well-defined variations with either latitude or season. Beyond the tropics, however, the detailed picture is much more complex. Here, there can be such large fluctuations in column ozone from day to day and year to year – and from place to place – that it becomes very difficult to detect any underlying 'signal' of a small consistent long-term downward trend that could be attributed to human activities. To do that, the basic requirement is for an extended, and reliable, dataset.

The longest data record of ozone measurements comes from the network of ground-based stations referred to in Figure 19 – the so-called *Dobson network*, of which Halley Bay is a part. This network was extended in 1957, but the distribution of stations remained uneven, with 42 out of 67 in northern temperate latitudes. Since 1979 the problem of spatial coverage had been much alleviated by instruments borne aloft on satellites, such as the Total Ozone Mapping Spectrometer, TOMS, on board NASA's Nimbus 7 satellite

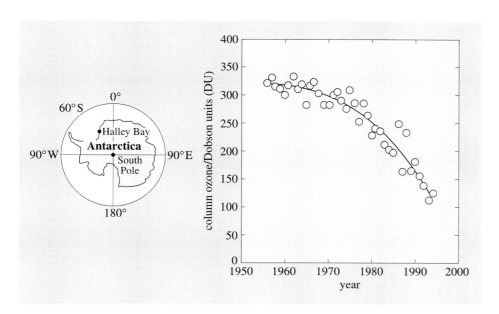

Figure 24   The October average ozone column measured from the BAS station at Halley Bay (76° S). Before the mid-1970s, values were around 300 DU, but since then a rapid decline has occurred. In the early 1990s, losses regularly exceeded ~50%.

(sometimes referred to as Nimbus-TOMS). But this data record was too short to permit the identification of any statistically significant trends. Overall, then, the early 1980s saw little convincing evidence that CFCs had *already* caused damage to global ozone levels.

5.1 Antarctica: the ozone hole

5.1.1 Observations

The results from Halley Bay have since been amply confirmed by other workers. Particularly telling are the computer-processed images from the TOMS data (Plate 1), which show that the depleted region now regularly extends over the entire Antarctic continent – and beyond. Other measurements, from satellites and balloons, have revealed that the depletion is mainly concentrated between about 12 and 24 km in altitude, spanning much of the lower stratosphere at these latitudes. A striking example from 1992 is shown in Figure 25.

The first hard evidence for a link between ozone loss and chlorine chemistry came from the Antarctic Airborne Ozone Experiment (AAOE), a major US-led experimental campaign during August and September 1987. The project involved some 150 scientists and technicians from several nations, based at Punta Areñas on the southern tip of Chile. Crucially, the team was equipped with a modified U2 spy plane (an ER-2), capable of flying *into* the depleted region, at altitudes up to 18 km. As indicated in Box 4 (and in the video sequence), it was instruments aboard the ER-2 that caught chlorine monoxide red-handed.

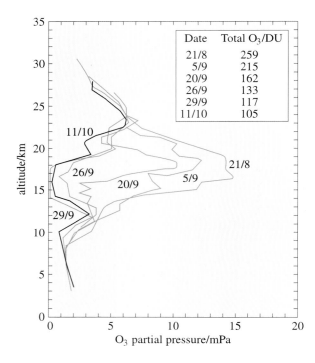

Date	Total O$_3$/DU
21/8	259
5/9	215
20/9	162
26/9	133
29/9	117
11/10	105

Figure 25 The development of the ozone hole over the South Pole between late August and early October 1992: notice that at some altitudes ozone was *completely* removed. Notice too that the horizontal axis records the partial pressure of ozone: this is directly proportional to its mixing ratio (recall SAQ 1).

The observations summarized above and in Box 4 serve to highlight several important issues that had to be addressed by the researchers involved:

- 'Standard' models of stratospheric ozone chemistry predict that most of the available chlorine should be locked up as HCl and ClONO$_2$ in the lower stratosphere (see Box 3 and the answer to SAQ 9), with only very small amounts of 'free' ClO·. Why, then, the dramatic increase in ClO· poleward of about 65° S?

- There is insufficient atomic oxygen in the low stratosphere (Section 3.2) for the 'standard' (X $\longrightarrow$ XO $\longrightarrow$ X) catalytic cycles to operate efficiently in the altitude region where ozone loss is most severe. What other, more efficient, cycle is at work?

- Why Antarctica in springtime (September/October in the southern hemisphere)?

- Why the rapid onset since the late 1970s?

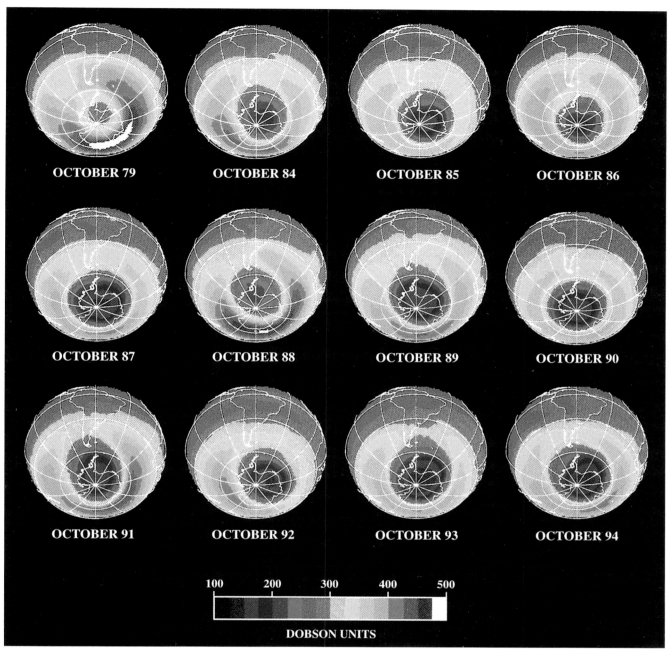

Plate 1 'Maps' of the ozone column (monthly mean total ozone) during October over the southern hemispere for 1979 and for 1984–1994, based on TOMS data. The picture for 1979 is typical of the situation that used to prevail over Antarctica, with column ozone at around 300–350 DU (green and yellow) throughout most of the winter and spring. Now ozone amounts over Antarctica (and beyond) fall rapidly with the return of sunlight in spring – usually to 150 DU or less (purple) – before recovering again during the summer. Through to the late 1980s, there seemed to be a roughly biennial cycle in the severity of the ozone hole (possibly linked to natural variations in the dynamics of the stratosphere). That pattern has since been broken – with six consecutive 'bad' years, 1989–1994: the ozone loss is both deeper and more extensive than it used to be, and the recovery to summer values takes longer.

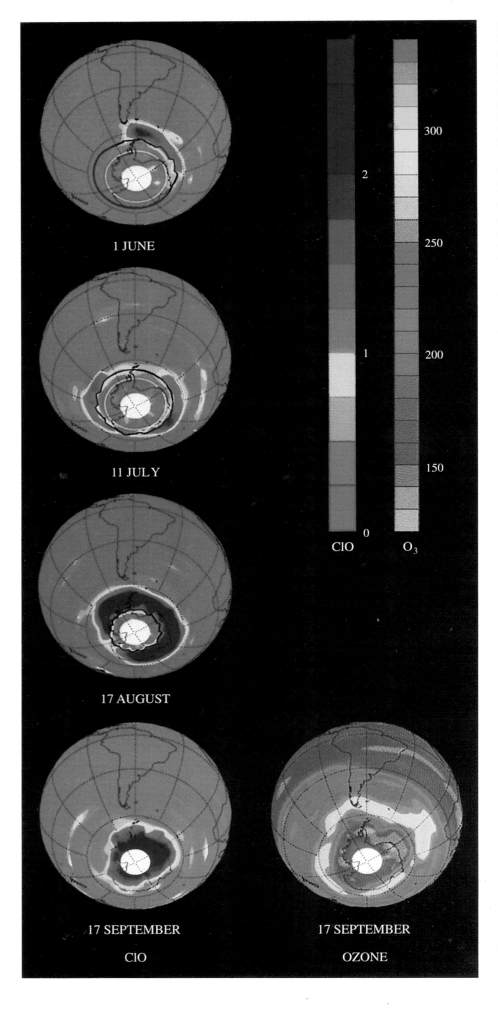

Plate 2 The MLS instrument on UARS detects microwave emissions, not only from ClO, but also from ozone (and several other species). Processed by computer, the data again provide global 'maps' of the species' abundance. Here, O_3 is colour-coded in Dobson units, whereas the ClO scale is 10^{19} molecules m^{-2} in a vertical column; 1 unit corresponds to about 1 p.p.b.v. of ClO in the low stratosphere. The data shown relate to the Antarctic winter of 1992: the black contour in the ClO maps marks the edge of daylight, poleward of which ClO is expected to be present as its dimer, Cl_2O_2. By the beginning of June, ClO was already at enhanced levels in the sunlit region at the edge of the Antarctic vortex – and it remained high all winter. With the polar cap fully sunlit later in the season, there was a striking correlation between high ClO (red and purple) and low ozone (blue and mauve).

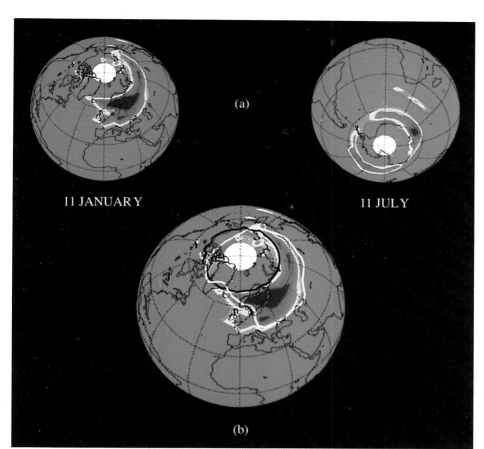

Plate 3 (a) On the same 'solar day' in 1992 – 11 January in the north and 11 July in the south – there is more ClO (left) in the north than in the south (right).

(b) On the same 11 January 1992 map of ClO as in (a), the black contour marks the edge of daylight. The green contour indicates where it was cold enough for polar stratospheric clouds to form – a condition ripe for the release of active chlorine radicals: notice that this sunlit portion coincides with the highest levels of ClO. The white contour indicates the approximate boundary of the Arctic vortex. The more disturbed circulation in the north shifts the vortex off the pole more than in the south – bringing air processed by PSCs into sunlight. On the other hand, the dynamics that distort the vortex also bring in warmer air from lower latitudes, which is the reason why the high levels of ClO don't stay around as long in the north, and that no Arctic ozone hole has (yet) appeared.

11 JANUARY

11 JULY

(a)

(b)

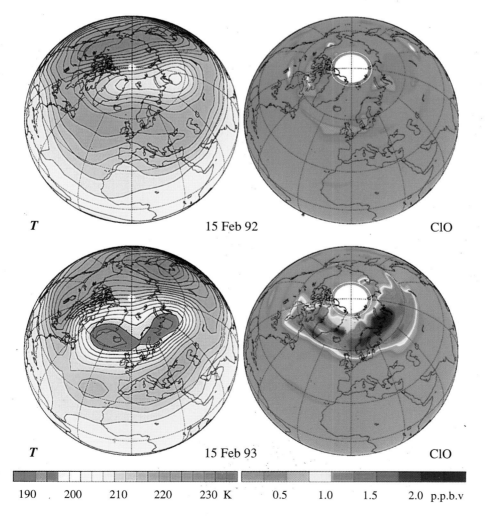

Plate 4 Temperatures in the lower stratosphere in winter are much more variable in the northern hemisphere than they are in the south. This comparison shows the striking difference in ClO abundance in the northern hemisphere on 15 February 1993, when temperatures were slightly below the threshold (~195 K) for PSCs to form, and on 15 February 1992, when temperatures were slightly above this threshold. Ozone was observed to be significantly lower in 1993 than in 1992, although this anomaly was not restricted to northern high latitudes (see Section 5.2.3).

T   15 Feb 92   ClO

T   15 Feb 93   ClO

190   200   210   220   230 K   0.5   1.0   1.5   2.0 p.p.b.v

How Polar Stratospheric Clouds Help Chlorine Destroy Ozone

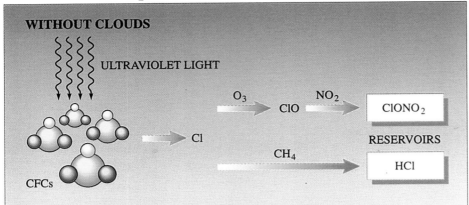

WITHOUT CLOUDS

Ultraviolet light from the sun breaks chlorofluorocarbons (CFCs) apart. The resulting chlorine (Cl) exists either as chlorine monoxide (ClO), formed in a reaction with ozone (O_3), or as free chlorine. Gases in the atmosphere, such as nitrogen dioxide (NO_2) and methane (CH_4), react with ClO and Cl to trap the chlorine in inert chemical reservoirs of chlorine nitrate ($ClONO_2$) and hydrochloric acid (HCl). Ozone depletion is minimal.

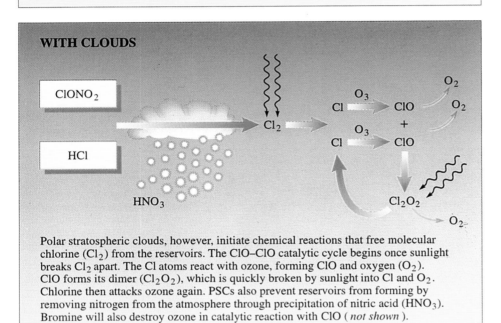

WITH CLOUDS

Polar stratospheric clouds, however, initiate chemical reactions that free molecular chlorine (Cl_2) from the reservoirs. The ClO–ClO catalytic cycle begins once sunlight breaks Cl_2 apart. The Cl atoms react with ozone, forming ClO and oxygen (O_2). ClO forms its dimer (Cl_2O_2), which is quickly broken by sunlight into Cl and O_2. Chlorine then attacks ozone again. PSCs also prevent reservoirs from forming by removing nitrogen from the atmosphere through precipitation of nitric acid (HNO_3). Bromine will also destroy ozone in catalytic reaction with ClO (*not shown*).

Plate 5   'How polar stratospheric clouds help chlorine destroy ozone'. Reproduced from O.B. Toon and R.P. Turco, 'Polar Stratospheric Clouds and Ozone Depletion', *Scientific American,* June 1991, pp. 40-47. (See the answer to Exercise 2.)

BOX 4

VITAL CLUES FROM THE ANTARCTIC AIRBORNE OZONE EXPERIMENT

Schematic [record] of observations from the ER-2 showed that at ~18 km altitude, poleward of ~65° S, the composition of the vortex [see Section 5.1.2] was highly perturbed. As this chemically perturbed region (CPR) was entered, the concentration of chlorine monoxide (ClO) increased sharply over several hundred kilometres, reaching values some 10 times greater than those observed immediately outside and some 100 times greater than those at lower latitudes. In late August ozone concentrations were roughly constant across this boundary.

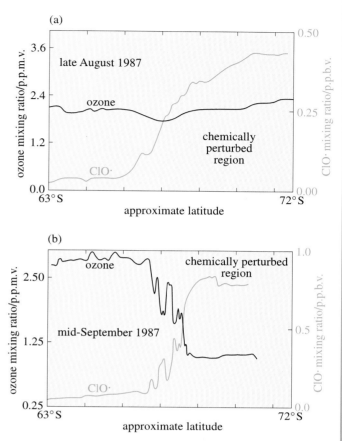

However, by the middle of September there was a sharp decline in ozone as the chemically perturbed region was entered. As well as the gross changes in ozone and ClO seen as aircraft flew into the CPR, there were also smaller-scale changes in ozone and ClO which [mirror one another]. This decline of ozone only in the region where the ClO concentrations were high provides a strong indication that chlorine chemistry is responsible for the ozone depletion.

5.1.2 What causes the ozone hole?

At the time of writing (1995), the consensus view is that a combination of meteorological and chemical factors is responsible for the peculiar efficiency with which chlorine destroys ozone over Antarctica. In outline, the main points are as follows.

At the autumn equinox (the end of March), the Sun sets for six months at the South Pole, and an area of darkness spreads out over the polar cap. Rapid cooling sets up a pattern of very strong westerly winds in the stratosphere – the so-called **polar vortex** (Figure 26), extending to around 60° S. There is a computer-generated visualization of this 3-D 'structure' in the video sequence. Based on actual meteorological data, this visualization reveals that the Antarctic vortex is extremely stable, enduring throughout the southern winter and spring (*after* the return of sunlight, that is): it does not finally break down until summer, usually sometime in November. While the vortex persists, polar air is largely sealed off from air at lower latitudes.

The unique meteorology of the polar stratosphere sets the scene for some hitherto 'unexpected' chemistry. Eventually the frigid core of the vortex becomes cold enough (temperatures lower than about 195 K, −78 °C, seem to be necessary) for icy particles to form – known as **polar stratospheric clouds**, or **PSCs**. Probably composed of solid phases of mixed nitric acid and water (see Box 5), PSC particles provide

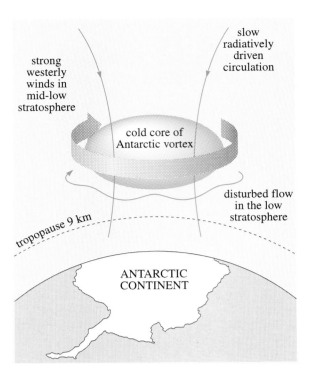

Figure 26 The Antarctic vortex forms as air cools and descends during the winter months. The circulation becomes westerly and strengthens, with winds in the low stratosphere reaching speeds of ~100 m s^{-1} (~225 miles per hour), or more, by the spring.

surfaces that promote fast *heterogeneous* reactions that involve one *or both* of the chlorine reservoirs (HCl and ClONO$_2$), reactions that are extremely slow in the gas phase. For example, laboratory studies indicate that the following reactions all have high 'reaction probabilities' (defined as the fraction of collisions that result in the immediate chemical conversion of the colliding species) on PSC-like surfaces:

$$ClONO_2(g) + H_2O(s) \longrightarrow HOCl(g) + HNO_3(s) \qquad (21)$$

$$HOCl(g) + HCl(s) \longrightarrow Cl_2(g) + H_2O(s) \qquad (22)$$

$$ClONO_2(g) + HCl(s) \longrightarrow Cl_2(g) + HNO_3(s) \qquad (23)$$

where HCl(s) represents HCl 'dissolved' into the particle surface: there is a little more about this in Box 5, *if you are interested*. Details apart, the point to note is that 'processing' by PSCs has two important consequences.

First, the heterogeneous reactions in equations 21 to 23 perturb the normal (i.e. gas-phase) partitioning of the chlorine derived from CFCs, releasing species (Cl$_2$ and HOCl) that are readily photolysed *even in low light conditions*. Photolysis generates Cl atoms – and hence ClO·, via reaction with O$_3$:

$$Cl· + O_3 \longrightarrow ClO· + O_2 \qquad (24)$$

Second, the nitric acid formed in the heterogeneous reactions above remains 'locked up' in the particles – whence the designation, HNO$_3$(s) – so that gas-phase concentrations of the main store of active nitrogen are reduced. There are now indications that heterogenous reactions involving N$_2$O$_5$ (the other, more temporary, reservoir of NO$_x$), such as:

$$N_2O_5(g) + H_2O(s) \longrightarrow 2HNO_3(s) \qquad (25)$$

may help this process of 'denoxification'. Sedimentation of cloud particles (Box 5 again) further depletes the store of nitrogen compounds in the polar stratosphere. This is important because it slows down the process that normally sequesters ClO· as chlorine nitrate (recall Figure 12 and the answer to SAQ 5):

$$ClO· + NO_2 + M \longrightarrow ClONO_2 + M \qquad (26)$$

═══ BOX 5 ═══

POLAR STRATOSPHERIC CLOUDS (PSCS) AND THE OZONE HOLE

Satellite data reveal that PSCs occur over Antarctica in winter when the temperature drops to about 195 K, which is several degrees *above* the threshold (the 'frost point', 188 K) for water-ice to form under the very dry conditions that prevail in the stratosphere (recall Box 1). As yet, the detailed composition of PSC particles is uncertain, and the 'microphysical' processes that govern their formation and growth are not well understood. However, on the basis of existing data, the essential features of their role in Antarctic ozone loss appear to be as follows.

1 Microphysical processes

As temperatures drop below 195 K, the co-condensation of nitric acid and water vapour produces *Type-I* PSC particles (see below), possibly in the crystalline form of nitric acid trihydrate (NAT), although there are indications that supercooled liquid droplets may be present as well. If further cooling takes the temperature below the frost point, these particles act as condensation nuclei for the growth of water-ice (*Type-II*) PSC particles. The latter can grow large enough (10–100 μm in diameter) to fall out of the stratosphere, through gravitational sedimentation.

2 Chemical processing by PSC particles

A PSC particle has yet to be brought back to the laboratory! But researchers have developed ways of producing suitable 'mimics' (commonly, thin films of water-ice or solid nitric acid hydrates), and have used various techniques to study their interactions with the chlorine reservoirs ($ClONO_2$ and HCl) under 'stratospheric' conditions. As indicated in the video sequence, a combination of infrared spectroscopy (to probe the surface species involved; you will learn more about this in Block 6) and mass spectrometry (to analyse the gases released) has proved to be particularly fruitful in providing clues to why the heterogeneous processes in equations 21 to 23 are so fast. In each case, a key step in the mechanism appears to be the reaction between an incoming chlorine species and a water molecule at the particle surface to produce *ionic* species, which may then react rapidly to release products. For example, the existing data support the following mechanistic scheme for reaction 21:

$$[H_2O] + ClONO_2(g) \longrightarrow [H_2OCl^+ + NO_3^-]$$
$$\xrightarrow{[H_2O]} HOCl(g) + [H_3O^+ + NO_3^-]$$

where the square brackets indicate surface species. Similarly, reaction 22 appears to involve formation of Cl^- at the surface, which reacts with incoming $HOCl$ to release Cl_2. Because reaction 23 is just the sum of reactions 21 and 22, the surface species H_2OCl^+ and Cl^- are likely to be implicated there as well.

Type-I cloud formation   Type-II cloud formation   sedimentation

In short, the net effect of PSC processing is to release Cl· from its reservoir molecules *and* to siphon off reactive nitrogen, reducing this sink for active chlorine. There is now good evidence that the resulting build-up in ClO· starts early in the season. The evidence comes from the Microwave Limb Sounder (MLS), one of ten remote sensors on board NASA's Upper Atmosphere Research Satellite (UARS). Launched by the space shuttle on 12 September 1991, MLS is providing the first global 'maps' of ClO· abundance – comparable with the computer-processed images from the TOMS ozone data. MLS measured more than 1 p.p.b.v. of ClO· in the Antarctic vortex in early June 1992 (see Plate 2).

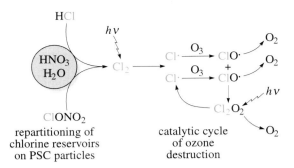

HCl

$h\nu$

O_3

O_2

O_2

Cl·

ClO·

+

Cl·

O_3

ClO·

$h\nu$

HNO₃
H₂O

Cl₂

ClONO₂

repartitioning of
chlorine reservoirs
on PSC particles

Cl₂O₂

catalytic cycle
of ozone
destruction

O_2

Figure 27 A simplified representation of the catalytic cycle implicated in the formation of the Antarctic ozone hole.

With O_3 present at a few parts per *million*, the reaction that forms ClO· (equation 24) cannot, *by itself*, cause significant ozone depletion. That requires some sort of catalytic cycle. As noted earlier, the step that completes the standard cycle, and hence regenerates Cl· (ClO· + O $\longrightarrow$ Cl· + O_2), is ineffective in the lower stratosphere. Instead, the most important cycle is believed to be the one involving the ClO· dimer, Cl_2O_2, shown schematically in Figure 27: spelt out in more formal terms, it reads:

$$ClO· + ClO· + M \longrightarrow Cl_2O_2 + M \tag{27}$$
$$Cl_2O_2 + h\nu \longrightarrow Cl· + ClO_2· \tag{28}$$
$$ClO_2· + M \longrightarrow Cl· + O_2 + M \tag{29}$$
$$2 \times (Cl· + O_3 \longrightarrow ClO· + O_2) \tag{24}$$

net: $\qquad 2O_3 \longrightarrow 3O_2$

Notice that the catalytic effect depends on the *photolysis* of the dimer (equation 28): thermal decomposition of this unstable species (the reverse of reaction 27) would simply regenerate ClO·, and hence shortcircuit the cycle. Thus, efficient operation of the cycle requires low temperatures and exposure to sunlight – just the conditions that prevail in the Antarctic stratosphere as the polar night retreats in spring. Calculations indicate that this cycle can account for most (~70%) of the observed ozone loss, with a smaller (possibly around 25%) contribution from a cycle that also involves bromine (recall Exercise 1 in Section 3.6):

$$ClO· + BrO· \longrightarrow Br· + Cl· + O_2 \tag{30}$$
$$Cl· + O_3 \longrightarrow ClO· + O_2 \tag{31}$$
$$Br· + O_3 \longrightarrow BrO· + O_2 \tag{32}$$

net: $\qquad 2O_3 \longrightarrow 3O_2$

Later in the season, 'normal' partitioning of the available chlorine is restored as rising temperatures evaporate PSCs, releasing HNO_3 – thus increasing the supply of NO_x, and hence sequestering ClO· as chlorine nitrate (via reaction 26). Final breakdown of the vortex in summer permits an influx of warm, ozone-rich air from lower latitudes, allowing the gradual recovery to 'summer' values evident in Figure 28.

In conclusion: the explanation summarized above effectively answers three of the four questions posed at the end of Section 5.1.1. Thus, the crucial ingredients for a major seasonal loss of ozone over Antarctica are a strong polar vortex, with a frigid core – well laced with PSCs – *that is maintained as long as possible after the return of sunlight*. Under these circumstances, heterogeneous processes can enhance the abundance of active chlorine in the low stratosphere, increasing its impact as an ozone sink, mainly via the catalytic cycle in Figure 27. These processes must have operated *before* the development of the ozone hole, but their influence was limited owing to the much smaller amount of chlorine in the stratosphere before the advent of CFCs. In which context, the record from Halley Bay (Figure 24) suggests strongly that chlorine levels passed a critical threshold in the late 1970s. The following SAQ invites you to think about this, and about the worsening situation since then (the last of the questions referred to above).

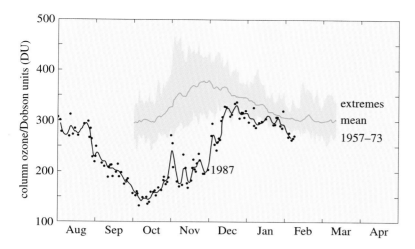

Figure 28   Up to the mid-1970s, column ozone values over Halley Bay remained roughly constant (at around 300 DU) through to mid-October, before rising with the influx of warm ozone-rich air from lower latitudes in summer. The recent pattern is very different, with column ozone falling rapidly in spring – often to less than 150 DU – before recovering again in the summer.

SAQ 10 (revision)   The chlorine loading of the atmosphere is thought to have increased from ~1.2 p.p.b.v. in 1970 to ~2.5 p.p.b.v. in 1980: by 1990, it had reached some 3.5 p.p.b.v. Assuming that formation of the ClO· dimer (equation 27) is the rate-limiting step in that catalytic cycle, write an expression for the rate of ozone destruction (i.e. $-d[O_3]/dt$) by this cycle. Can you now suggest one factor that may go some way to explaining the sudden appearance – and increasing severity – of the ozone hole?

The models discussed in Section 4 did not contain any of the special ingredients outlined in this Section: none of them incorporated any heterogeneous chemistry, and even the best 2-D model cannot fully simulate the complex meteorology of the polar vortex. *With the benefit of hindsight*, the fact that such models failed to predict the ozone hole takes on a certain inevitability.

5.2   Global ozone trends

With the recognition that heterogeneous processes are deeply implicated in the formation of the Antarctic ozone hole, a broader set of questions arises. Are similar processes at work elsewhere in the stratosphere? And if so, what is the prognosis for the ozone shield over the Arctic – and potentially far more important, over the heavily populated regions at lower latitudes as well?

5.2.1   Observations

The questions posed above take on a sharper focus in the light of evidence that the world *has* now seen a real depletion of column ozone at all latitudes outside of the tropics. There were indications to this effect in the report from an international panel of scientists, the Ozone Trends Panel, published in 1988. Their key finding was that a thorough re-evaluation of the ground-based data – analysed this time over specified latitude bands and by season, not globally and annually as it had been before – did show statistically significant long-term downward trends over the period 1969–1986. The picture that emerges from these, and more recent, ground-based measurements is broadly consistent with that revealed by analysis of the ozone data collected by Nimbus-TOMS (Figure 29): almost fifteen years of global data were obtained before the instrument ceased to function in May 1993. A similar instrument on a Russian satellite (Meteor-TOMS) has taken over since then.

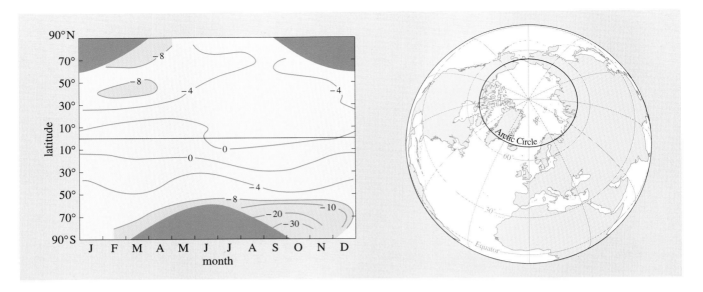

Figure 29   Total ozone trends shown as percentage change per decade, deduced from Nimbus-TOMS data for the period 1979–1990, analysed according to latitude and season. The coloured areas have rates of loss exceeding 8% per decade. The dark green areas show the Arctic and Antarctic polar nights, in which TOMS is unable to make measurements.

There are two main points to take from Figure 29:

● There are large losses at the highest southern latitudes, associated with the recurrence of the Antarctic ozone hole during the austral (southern) spring: the peak downward trend is more than 35% per decade.

● There is no evidence of an ozone hole, as such, over the Arctic. However, peak losses at northern *mid*-latitudes (30–60° N) in winter and early spring are over 8% per decade, and *larger* than those at comparable southern latitudes. Or to put it more sharply, greater springtime losses occurred between 1979 and 1990 above the UK and Europe than above, for example, South America or New Zealand – despite their proximity to the Antarctic ozone hole.

Averaged over the whole globe (or strictly, between 65° S and 65° N, because there are no data in the polar night), Nimbus-TOMS measurements show a downward trend in column ozone of about 3% per decade between 1979 and the end of 1990. The decline continued at much the same rate during 1991, but thereafter the ozone loss *accelerated*. As shown in Figure 30, the eighteen months through to mid-1993 saw global average ozone fall to record low values, with indications of a slow recovery beginning in the second half of 1993. The decline during winter at northern mid-latitudes was even more dramatic than the global average: here, the early months of 1993 saw column ozone values some 10–15% below the mean for 1979–1990.

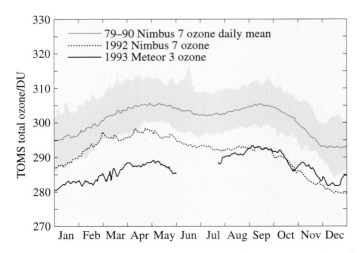

Figure 30   TOMS measurements of the global mean ozone column for 1992 and 1993, contrasted with the daily mean value (green line) and range (dark green tone) for 1979–1990. The 1993 values are Meteor-TOMS measurements; the remaining data come from the Nimbus-TOMS record.

Other ground-based and satellite data reveal that the bulk of the ozone loss at all latitudes has occurred in the *lower* stratosphere (just as it has over Antarctica), although there are indications of some depletion at around 40 km as well. Recall that this is the region where stratospheric ozone was predicted to be most vulnerable to enhanced levels of chlorine according to the 'gas-phase' chemistry discussed in Section 3 – a pattern evident in the modelling studies recorded in Figure 23a.

The mechanisms responsible for the trends outlined above are very much the subject of ongoing research. Current thinking has drawn not only on continued satellite surveillance and ground-based measurements, but also on the vast amount of data collected by a series of intensive field campaigns (comparable to AAOE), two of which are still underway at the time of writing. Acronyms abound! Those of some of the major campaigns to date are collected in Box 6, together with an indication of their timing, and of the latitudes subjected to intensive study. As the measurements made during these experimental programmes are analysed and appear in the literature, the complexity of the ozone budget in the stratosphere becomes ever more apparent. A recurrent theme is the importance of both chemical and meteorological factors, and of the interplay between them.

BOX 6

PROBING STRATOSPHERIC OZONE

The discovery of the Antarctic ozone hole sharpened concern throughout the world about the stability of the stratospheric ozone layer. That concern has, in recent years, prompted a series of intensive field campaigns in both hemispheres. Over the period 1987–1995, these have included three further US-led programmes (again involving the ER-2) – two to the north (Airborne Arctic Stratospheric Expeditions, AASE), and one to the south (Airborne Southern Hemisphere Ozone Experiment, ASHOE) – together with two major European projects (the European Arctic Stratospheric Ozone Experiment, EASOE, and the Second European Stratospheric Arctic and Mid-latitude Experiment, SESAME).

Of the campaigns included in the Figure, SESAME is devoted to a prolonged study of the processes occurring in – and connecting – the lower stratosphere at high and middle northern latitudes: it is designed to run throughout 1994 and 1995. The overall strategy calls for a large number of different, but complementary, measurements at sites scattered in and around the Arctic and Europe, covering an area from Greenland to northern Russia and stretching as far south as Greece. Aspects of this campaign feature in the video sequence associated with this Topic Study.

(See Section 5.2.3 for the significance of the 'sulfate aerosol layer'.)

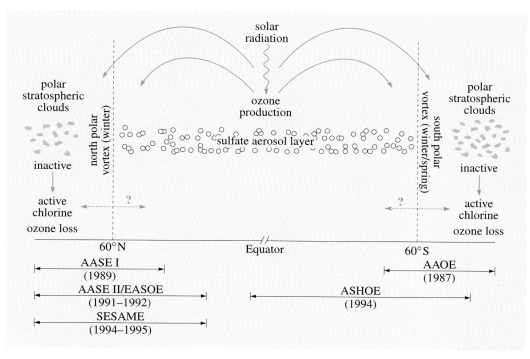

5.2.2 Northern polar regions: why no Arctic ozone hole?

For reasons that are linked to the different distribution of landmasses and oceans in the two hemispheres, the winter circulation of the stratosphere is generally more disturbed in north polar regions than in the south. This has several important consequences. Minimum temperatures in the Arctic lower stratosphere tend to be some 5–10 K warmer than over Antarctica (see Figure 31), PSCs are less abundant, and the circumpolar vortex is weaker and more distorted, and breaks down earlier in the season than its southern counterpart. In short, the Arctic vortex and the PSCs within it rarely persist until northern polar regions are fully sunlit.

- Why would this affect the severity of ozone depletion over the Arctic?

- Catalytic ozone removal on the scale that occurs over Antarctica requires that PSCs persist as long as possible *after* the return of sunlight, thereby maintaining enhanced levels of active chlorine.

As noted earlier, nothing comparable with the Antarctic ozone hole has, as yet, been detected over the North Pole in spring. However, there *is* good evidence that the chemical composition of the Arctic stratosphere in winter can be as highly perturbed as that in Antarctica. Northern polar regions sampled during the aircraft campaigns of the 1988/89 and 1991/92 winters showed the characteristic 'fingerprint' of air that has been processed by PSCs – dramatically enhanced levels of $ClO\cdot$. And so do the data collected by the MLS. A striking example from early January 1992 is shown in Plate 3: at this time, $ClO\cdot$ concentrations in the Arctic vortex were as high as any ever observed in its southern counterpart.

In short, the Arctic stratosphere is primed for chlorine-catalysed ozone loss. But to what extent is that actually happening? This turns out to be a very tricky question to answer with any confidence. The disturbed winter circulation of the northern stratosphere means that transport by air motions can itself cause changes in the ozone concentration at a given location and altitude. Given ozone measurements – x days apart, say – it is difficult to tease out the 'signal' of *chemical* destruction from the 'noise' generated by such *dynamic* fluctuations. Atmospheric scientists have devised several ingenious techniques for getting a handle on this problem. Suffice it to say that they use different sets of observations and methodologies to derive ozone losses, but come up with broadly consistent results. As far as 1992 is concerned, for example, these analyses indicate that *chemical* processes removed about 15–25% of the ozone at altitudes near 18 km during January, the month with the strongest perturbation in chlorine chemistry.

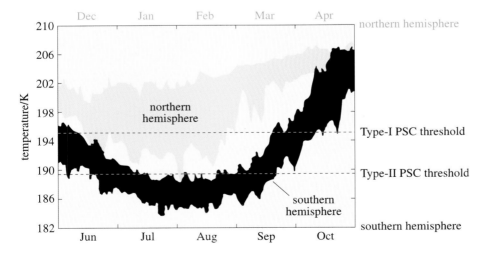

Figure 31   Range of winter/spring temperatures for equivalent months in the polar vortex of the northern and southern hemispheres for the period 1980–1988, deduced from satellite measurements of emission in the microwave region. Notice the greater variability in the north, with temperatures dipping below the threshold for PSC formation for only a month or so in winter. By contrast, PSCs can persist well into spring over Antarctica.

More generally, all of the studies referred to above have served to underline the importance of the meteorological conditions that prevail in any given year. Thus, for example, the extremely high ClO· levels in the Arctic vortex did not last long in 1992. Rapid warming and mixing of air masses during late January drastically reduced the amount of active chlorine – and the region was spared a deeper, and more sustained, ozone loss. But with temperatures just a few degrees colder in February 1993, ClO· remained enhanced through to the end of the month, and ozone concentrations were significantly lower than in February 1992 (see Plate 4). The important general point is that the 'preconditioning' of the Arctic stratosphere is likely to be a recurring feature of the northern winter months – or at least, it is until the total chlorine loading of the atmosphere starts to fall. Under these circumstances, major ozone loss over northern high latitudes could be triggered by an unusually prolonged period of very cold winter temperatures in the low stratosphere, and records suggest that this does occur in some years. One recent example is referred to in the video sequence.

5.2.3 What about the trends at mid-latitudes?

These trends are unexpectedly large – or to be more precise, the ozone depletion at mid-latitudes is more severe than that predicted by models based on gas-phase chemistry *alone*. Although a detailed explanation for these trends has yet to emerge, there is a list of plausible contributing factors. In one way or another, all are linked to the influence of heterogeneous chemistry: they fall into two broad categories.

The influence of polar processes

It is generally agreed that ozone-poor air is transported away from the pole during the final breakdown of the Antarctic vortex, thus effectively diluting the ozone column at lower latitudes in the southern hemisphere. More controversial is the suggestion by some researchers that the vortex is not as 'isolated' as previously supposed. In essence, the idea is that it could act like a 'flowing processor', with PSCs within the vortex thus capable of processing a large volume of air, which is then transported out to *sunlit* regions at lower latitudes. The impact this might have at mid-latitudes – as a source of air that is poor in ozone and rich in ClO· – depends critically on the details of mass exchange across the vortex edge. And this is not well understood at present: indeed, it remains a subject of intense and lively debate among atmospheric dynamicists!

The influence of sulfate aerosols

Clues to another factor that may be implicated in the trends at mid-latitudes have come from observations of ozone depletion in the wake of major volcanic eruptions. Thus, the low column ozone at northern mid-latitudes noted earlier followed the eruption of Mount Pinatubo (15° N) in the Philippines in June 1991. Although the strength and duration of these recent anomalies are the largest on record, the Pinatubo eruption is not without precedent in affecting the ozone layer. Column ozone fell to then-record lows at many Dobson stations in northern mid-latitudes during the winter following the eruption of El Chichón in Mexico in early 1982.

The timing of these events has focused interest on the possible chemical effects of a fine mist (or *aerosol*) of droplets of sulfuric acid that pervades the lower stratosphere. The normal background level of this global **sulfate aerosol layer** is maintained by a steady supply of gases transported up from the troposphere, the most important of which is carbonyl sulfide (COS): photolysis of this releases atomic sulfur, which then undergoes a stepwise oxidation via SO_2 to sulfuric acid. However, major volcanic eruptions, such as those of El Chichón and Mt Pinatubo, inject huge amounts of SO_2 *directly* into the low stratosphere: as a result, aerosol concentrations can increase by a factor of as much as a 100. For example, observations after the Pinatubo eruption tracked the volcanic cloud as it spread around the globe: it produced a peak enhancement (by a factor of 30–50) in the aerosol load at northern mid-latitudes in April 1992. By early 1994, the volcanic aerosol was beginning to fade away, and ozone amounts were recovering to more normal values.

Laboratory studies have shown that sulfate aerosols do indeed support heterogeneous reactions that can lead to elevated levels of ClO·, either *directly* (via reaction 21 for example, which frees active chlorine from the $ClONO_2$ reservoir) or *indirectly* (via reaction 25, which effectively siphons off reactive nitrogen). But the rates of these processes are highly dependent on the degree of hydration of the sulfuric acid droplets, which is itself a strong function of temperature: the colder it is, the greater the water content of the aerosol and the faster the reactions proceed. The existing data suggest that these processes are unlikely to have much impact on the ozone budget at middle – or indeed, lower and warmer – latitudes. On the other hand, there is some evidence that heterogeneous processing by sulfate aerosols was occurring at somewhat higher (and colder) latitudes in the aftermath of the Pinatubo eruption. Some of the air masses sampled during AASE II (winter of 1991/92, see Box 6) did show a correlation between enhanced (volcanic) aerosol and perturbed chemistry (i.e. low NO_x and high ClO·). However, it is not clear whether that made a significant contribution to the severe perturbation of the ozone record in 1992 and 1993. The role in chlorine activation played by *background* levels of aerosol has yet to be established.

5.3 Drawing the threads together

The history of research into CFC-linked ozone depletion has been one of recurring surprises. The first shock came with the discovery of the ozone hole over Antarctica. There was another surprise in 1989, when the first comprehensive field campaign to probe northern polar regions (AASE I, see Box 6) produced clear evidence of chemical processing by PSCs: before then, it was thought that chlorine activation on the scale observed could occur only during the frigid Antarctic winter. More recently, the TOMS data have revealed an unexpectedly large – if more gradual – thinning of the ozone shield over large tracts of North America, Europe and Asia. The record lows in 1992 and 1993 have added to the puzzle.

As you have seen, a major research effort has been devoted to unravelling the various strands of this puzzle. Satellite surveillance provides near-global coverage. Major field campaigns offer vital snapshots of the detailed chemical composition, temperature, PSC and aerosol content, and so on of different air masses in the sampled regions. Behind the scenes – but equally important – are laboratory studies of the heterogeneous processes that are believed to be at work in the stratosphere.

Melding these disparate elements together calls for the close involvement of the atmospheric modelling community. Here, recent years have seen progress toward the kind of 'ideal' 3-D chemistry-transport model envisaged in Section 4.2. As yet, computational constraints still preclude the *simultaneous* treatment of the full panoply of *both* (photo)chemical *and* meteorological processes. The development of such 'fully coupled' 3-D models must await a new generation of larger and faster supercomputers. In the meantime, one fruitful strategy has been to use a so-called 'decoupled' or 'off-line' 3-D model. Put simply, this approach uses meteorological parameters (wind speed, pressure, temperature, etc.) calculated by a GCM as *input* to a 3-D chemistry model, which then computes the time-evolving chemical composition for a few days. Used in this way, models that include the influence of heterogeneous processes can simulate the perturbed chemistry of the polar vortex. Or at least, they can reproduce the essential features of this seasonal phenomenon, indicating that the processes involved in polar regions are now reasonably well understood. There is an example of a simulation like this in the accompanying video sequence.

The task of disentangling the influence of chemical and dynamic factors elsewhere in the stratosphere is proving to be far more difficult. To condense the discussion in Section 5.2.3, the downward trend at middle latitudes may be caused by transport of ozone-poor air from polar regions. Or it may be due to *in situ* chlorine-catalysed ozone loss, *either* by transport of air rich in ClO· across the edge of the polar vortex, *or* (and this remains a possibility) by chlorine activation on sulfate aerosols. Whether the field campaigns that are underway at the time of writing will help to clarify the role of these various processes remains to be seen. However, one telling point is becoming apparent. Especially in the northern hemisphere, 'activated' air can be 'stripped off' the edge of the polar vortex during the winter months, either as a 'blob' or in a fine filament. Detailed study of these small-scale features – and more importantly, of their influence on the ozone budget at middle latitudes – will probably require 3-D models that have a higher spatial resolution than most existing versions, where the horizontal grid (see Figure 21 in Section 4.2) is typically about 6° latitude × 6° longitude. One further area of uncertainty concerns the part played by *bromine-catalysed* ozone loss, not only in polar regions (via the cycle in equations 30–32, for example) but possibly elsewhere as well. And given the surprises that litter this unfolding story, researchers are conscious that other, presently unsuspected, mechanisms may also be at work.

STUDY COMMENT   Instead of providing a summary at this point, I'd like to encourage you to do that for yourself. To that end, working through the following SAQ and Exercise should help you to check that you have both noted the main points in Section 5, and understood their implications as far as CFC-linked ozone depletion is concerned. Exercise 2 also gives you a chance to practise an important general skill – that of extracting the relevant information from one source (here, the text of this Topic Study), and then using it to comment on views expressed in another.

SAQ I I   The modelling studies discussed in Section 4.3 indicated that the build-up of CO_2 in the atmosphere could act to partially offset chlorine-catalysed ozone loss in the stratosphere. Given the scientific findings discussed in this Section, explain *briefly* why more recent assessments have questioned this view, suggesting instead that increased CO_2 could lead to *enhanced* ozone loss – especially in the lower stratosphere.

EXERCISE 2   In January 1992, during the AASE II field campaign, instruments aboard NASA's ER-2 recorded ClO· at a mixing ratio of 1.5 p.p.b.v. in the stratosphere north of Bangor, Maine, in the USA. The scientists involved interrupted their experimental flights to announce their findings – pointing out that, under these conditions, ozone could be destroyed at a rate of 1 to 2% a day. This press conference generated a storm of publicity in the USA, with media reports claiming that a northern ozone hole was imminent. Within days, then-President Bush moved unilaterally to tighten the US deadline for phasing out CFCs from 2000 to 1996. The non-appearance of an Arctic hole then provoked criticism that NASA's 'alarmist warning' was 'setting science policy by press release'. Other commentators went further, claiming that the scientific basis for banning CFCs was 'flimsy and dubious'.

Suppose that you have been asked to write a short *illustrated* article (*not more than a thousand words*) in response to comments like these, the aim being to set the events that provoked them in the wider context of the scientific case for phasing out CFC usage as soon as possible. Drawing on material in this and previous Sections, make a list of the points and diagrams that you would include in your article. Assume that it will be published in a popular periodical, like *New Scientist* or *Scientific American*.

Use the following notes as a guide:

1   Start with a 'punchy' opening that deals with the observed ozone losses.

2   Devote the bulk of the article to the arguments for a link between CFCs and polar ozone loss, including a *brief* (and inevitably much simplified) account of the chemistry involved. Try to get across the important differences, *and similarities*, between the two polar regions. Include a comment on the implications for ozone at middle latitudes.

3   Finish with a comment about 'future prospects', designed to underline the arguments for an early phase-out of CFCs.

6 WHAT MIGHT OZONE LOSS MEAN IN PRACTICE?

Concern about the possible consequences of stratospheric ozone depletion can be summarized as two broad questions. First, how would the resulting increase in uv radiation at the Earth's surface affect biological systems? And second, could changes in the amount and/or distribution of stratospheric ozone influence the world's climate?

STUDY COMMENT   There is not enough space here to explore these questions in detail, and anyway that would not be appropriate in a course primarily concerned with physical chemistry. However, we felt it important to set the science discussed earlier in the context of the potential consequences of sustained ozone depletion. A quick read through should suffice.

6.1 Biological effects of enhanced uv radiation at the surface

As noted earlier (see Figure 4), the shorter and most dangerous wavelengths in the incoming solar radiation, those in the band known as UV-C (less than 280 nm), never reach the Earth's surface: a small fraction of the oxygen and ozone present in the atmosphere is sufficient to maintain this protection. Equally, ozone depletion is of little consequence as far as UV-A (320–400 nm) is concerned: radiation at these wavelengths is only weakly absorbed by ozone, and anyway UV-A is relatively (though not entirely) harmless. By contrast, partial absorption by ozone greatly attentuates the solar input in the band known as UV-B (280–320 nm), but does not block it completely. Even the relatively low levels of UV-B that normally get through to ground level can have harmful effects, causing sunburn, eye damage and skin cancer in humans, for example. More generally, many biological systems are particularly sensitive to UV-B radiation, with the shorter wavelengths in this band being especially effective in causing damage to DNA and to the photosynthetic mechanism in plants.

The total amount of UV-B received at the surface should be highly sensitive to changes in the ozone column, an increase of around 1.5% for each 1% loss of ozone being the generally accepted figure. Further, ozone absorbs shorter UV-B wavelengths more effectively than longer ones (recall Figure 5). Thus, depletion should allow progressively more of the shorter – *and potentially more damaging* – wavelengths to reach the surface. With the notable exception of Antarctica, there is as yet little hard evidence that the real downward trends in ozone noted in the previous Section have produced *detectable* increases in UV-B. Here, the underlying problem is a familiar one: the ambient uv at the surface is naturally very variable. For a start, there is a strong latitudinal gradient: the intensity of UV-B is greater near the Equator, both because, on average, less ozone is found there (recall Figure 19), and because the Sun is more nearly overhead, so the radiation has a shorter path through the atmosphere. Equally, daily and seasonal variations in the elevation of the Sun influence the UV-B received at any given location – markedly so at higher latitudes – and so do natural fluctuations in the local ozone column and cloud cover. Tropospheric pollution also comes into the equation: aerosols and other particulate matter (smoke, dust, etc.) generally act to scatter and diminish incoming uv radiation.

Detecting the signal of a gradual trend in UV-B – above the noise of natural variability, that is – again calls for long-term data records. Unfortunately, most such records come from stations that use 'broad band' instruments. These are designed to record the *total* uv intensity over a broad spectral range between 280 and 400 nm: such measurements are dominated by the more intense longer wavelength UV-B and UV-A radiation, not by the shorter wavelengths most affected by ozone depletion. Spectrometers capable of collecting detailed UV-B spectral data have been in routine use at a few sites: but as yet, the records are too short to extract statistically significant trend information. On the other hand, instruments at several sites in Antarctica *have* detected the expected local increase in UV-B under conditions of severe ozone depletion in springtime.

With only limited spectral data currently available, researchers fall back on calculated estimates of the wavelength-dependent increase in UV-B that should accompany a given (observed or projected) level of ozone depletion. Using such estimates to assess the likely biological consequences is a difficult and uncertain business: few of the effects are sufficiently well understood at present for the impact of enhanced UV-B to be quantified.

6.1.1 Human health

The link between UV-B exposure and skin cancer is particularly emotive: here, there are two main strands of evidence. First, skin cancer is predominantly a disease of pale-skinned people, and the dark pigment – melanin – is known to be an effective filter of UV-B. The second strand comes from *epidemiology*, a study of the factors that influence the occurrence of the disease in human populations. Such studies reveal that the incidence of *non-melanoma* skin cancers (the most common form of the disease) is closely correlated with geographical latitude and with long-term UV-B exposure: the incidence is higher in regions of the world that experience a more intense uv 'climate', and it occurs predominantly on light-exposed areas of the skin, in the elderly and in those with outdoor occupations (farmers and sailors, for example). These cancers are distressing, but can usually be treated successfully.

Malignant melanocyte tumours (*melanomas*) are much rarer, but more serious. With a high risk of the disease spreading to other parts of the body (metastasis), the prognosis is poor: some 25% of patients die from the disease within five years. Unlike other skin cancers, the incidence of melanoma has been rising among pale-skinned races throughout the world. And the link with UV-B exposure is more complex: the disease tends to affect relatively young people, and does not show the same propensity to develop on skin regularly exposed to sunlight. Instead, the available data suggest that intermittent, *but severe*, exposures – sunburns, in short – are an important factor, especially if experienced early in life (before the ages of 10–14).

Epidemiological data, together with the results of experiments on animal 'models' (see Figure 32, for example), have been used as the basis for estimates of how ozone loss, and the attendant increase in UV-B, may affect rates of skin cancer. For instance, one recent (if still highly uncertain) estimate suggests that every 1% decrease in column ozone could result in a 3% rise in the incidence of non-melanoma skin cancers – which translates into some 12 000 to 15 000 extra cases a year in the USA – together with a possible 1% increase in mortality from melanoma. To put these figures in context, melanomas now kill an estimated 6 000 people a year in the USA: non-melanoma cancers, albeit some 20 to 30 times more common, account for roughly the same number of deaths.

Projecting the *global* toll in increased rates of skin cancer is even more problematic, partly because susceptibility to the disease is so dependent on skin type. On top of this, media coverage of the ozone issue has undoubtedly increased public awareness of the *current* risks of contracting skin cancer and of the sensible precautions that can be taken: reducing skin exposure on 'sunshine holidays', wearing protective sun-screens, etc. It is difficult to predict how such changes in personal behaviour may influence long-term trends.

Exposure to enhanced levels of UV-B can also have other directly harmful effects on human health, the two most serious being tendencies to suppress the body's immune responses and to cause eye damage (e.g. the development of cataracts). Although even more difficult to quantify, it is noteworthy that these effects would touch all populations, with some consequences – possible increases in the incidence or severity of infectious diseases, for example – likely to be particularly severe for people in developing countries.

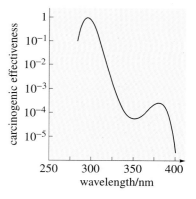

Figure 32 The relative effectiveness of different ultrviolet wavelengths in inducing skin cancer in a particular strain of hairless mice. Plots like this are known as *biological action spectra*: they serve as 'weighting functions', which can be used to assess the biological effectiveness of the change in UV-B resulting from a given ozone loss. In practice, the experiments required to prepare action spectra are difficult and time-consuming: few are available as yet.

6.1.2 Terrestrial plants

Virtually all life on Earth is ultimately dependent on plant photosynthesis. Through long evolution, the vegetation found in different regions of the world is adapted to the levels of radiation that it currently experiences. Information about the way plants respond to enhanced UV-B has come largely from experimental work carried out under greenhouse conditions, using uv-lamps to simulate various ozone depletion scenarios. To date, most studies have focused on agricultural crops typical of mid-latitudes. Of the 300 or so species and cultivars screened for tolerance to UV-B, some 60% have been found to be sensitive – although the degree of sensitivity varies widely, even among cultivars within a given crop species. Typically, sensitive plants show reduced growth and smaller leaves: unable to photosynthesize as efficiently as other plants, they yield smaller amounts of seeds or fruit. In some cases, these plants also show changes in their chemical composition, which can affect food quality and palatability.

Several commercially important conifer species also appear to be sensitive to UV-B (Figure 33), which could affect forest productivity. Potentially more important, it is possible that differing responses to increased UV-B levels could upset the competitive balance in natural ecosystems. Without long-term field studies, however, the impact that this might have is largely a matter of speculation – as, indeed, are the consequences for food production and forestry. In this context, research has highlighted another complication: uv effects can vary with changes in other environmental factors (notably temperature, and nutrient and water availability) – and with the ambient level of atmospheric CO_2. At this point, concern about ozone depletion becomes interwoven with that about the possible climatic consequences of the ongoing accumulation of CO_2 (and other greenhouse gases) in the atmosphere.

Figure 33   Loblolly pine trees irradiated with UV-B simulating 0%, 20% and 40% ozone depletion.

6.1.3 Aquatic ecosystems

Life in the oceans is also vulnerable to uv radiation. Although not as important as visible light or temperature or nutrient levels, there is evidence that ambient UV-B is nevertheless an important limiting factor in marine ecosystems. In practice, UV-B is rapidly attenuated in seawater, but it can penetrate to depths of about 20 m in clear waters. Evidence that ozone depletion is *already* affecting organisms that live close to the sea surface has come from the oceans around Antarctica – regions that support some of the largest communities of phytoplankton in the world. There, researchers have confirmed that the ozone hole has a seriously detrimental effect on the productivity of these microscopic plants: photosynthesis can be reduced by as much as 25%. Other studies have shown that enhanced UV-B can damage zooplankton (microscopic organisms that 'graze' the phytoplankton), crabs, shrimp and fish – especially during their vulnerable developmental stages, when their eggs and larvae occur at or near the ocean surface.

Damage to these small organisms is of particular concern because they are at the base of the marine food chain. Any changes here (in species composition, say) could have repercussions higher up, affecting all feeders – fish, birds and mammals included. And this could affect global food supplies: in 1990, roughly a third of the world's animal protein for human consumption came from the sea. So once again, the potential exists for a substantial *indirect* effect on human health. As with terrestrial ecosystems, however, too little is known for predictions of the overall biological consequences of ozone depletion to be made with any degree of confidence. Many uncertainties remain regarding the magnitude of the effects noted above: their implications for aquatic ecosystems on regional and global scales are unclear.

6.2 Effects linked to ozone's role in the climate system

Recent years have seen a growing awareness of important linkages and feedbacks between ozone change and climate, issues that were touched on in Section 4.3 (see SAQ 8), and explored further in SAQ 11 (Section 5.3). There we noted that a cooling of the stratosphere due to enhanced concentrations of CO_2 could act back on ozone chemistry. As was stressed in the answer to SAQ 11, this 'temperature-feedback' could act either way. Specifically, it could help to mitigate ozone loss in the middle to upper stratosphere (via gas-phase chemistry), *but it could well have the reverse effect lower down* – with increased chlorine-activation by PSCs (and, possibly, sulfate aerosols) leading to enhanced ozone losses.

In practice, recent decades have seen a cooling of the lower stratosphere (Figure 34). However, modelling studies indicate that these temperature trends *cannot* be attributed to the observed build-up of CO_2: the expected rate of this cooling is too slow. Rather, the predominant factor appears to be the way that ozone depletion *itself* feeds back on stratospheric temperature. Recall that ozone acts to warm the stratosphere: thus depletion of the ozone layer would be expected to have a cooling effect. Calculations suggest that the temperature trends in Figure 34 are broadly consistent with the observed ozone losses in the lower stratosphere. Notice in particular the marked cooling at high southern latitudes, reflecting the influence of the Antarctic ozone hole.

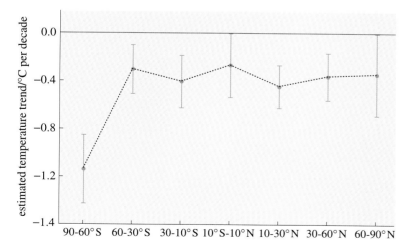

Figure 34   The dotted line connects estimates of the trend in lower stratospheric temperature (in °C per decade) for specified latitude bands over the period 1968–1989. The estimates are zonally averaged figures based on an analysis of data collected by balloon-borne instruments.

Looking to the future, the fact that PSC occurrence is determined by a threshold temperature means that a continued cooling of the lower stratosphere – whether induced by a growth in atmospheric CO_2, or by ozone depletion, or both – could trigger the feedback loop noted above. And that could increase the risk of an Arctic ozone hole – or, indeed, of deeper losses elsewhere.

One final point. The links between ozone and climate are not restricted to conditions within the stratosphere itself. In particular, it is now recognized that depletion of the ozone layer can induce a climate response at the Earth's surface – only this time, there are two opposing effects:

●   If more uv radiation gets through to the surface and lower atmosphere, that should contribute to surface warming;

●   On the other hand, a cooler stratosphere would be expected to emit *less* infrared radiation back down into the troposphere, which would be a cooling effect.

Based on the observed losses from the lower stratosphere, recent calculations suggest that the cooling effect is likely to dominate.

6.3   Summary of Section 6

1   In the absence of other changes, ozone loss should lead to enhanced UV-B at the Earth's surface, and shift the spectral distribution to shorter wavelengths. The expected increase in UV-B has been detected underneath the Antarctic ozone hole. Elsewhere, discerning a gradual trend above the noise of natural variability is hampered by the lack of UV-B data records.

2   Although the overall biological consequences of enhanced UV-B are uncertain, awareness is growing of the types of damage that might ensue – to human health and food supplies, and to terrestrial and aquatic ecosystems. The ozone hole is already affecting the productivity of phytoplankton in the waters around Antarctica.

3   A further area of concern stems from the fact that ozone is an important gas in the climate system: it can both cause and respond to changes in atmospheric temperature. Specifically:

●   Recent cooling of the lower stratosphere is probably due to ozone loss;

●   Whatever the cause (be it more CO_2, or less O_3, or both), a continued cooling of the lower stratosphere could enhance ozone losses, in general, and increase the risk of an Arctic ozone hole, in particular;

●   Changes to stratospheric ozone can also disturb the climate at the Earth's surface.

SAQ 12   As noted in the text, the ambient UV-B at the surface varies widely, with time of day, season and latitude. If there is no incident uv (at night) or if it is of very low intensity (high latitudes in winter), then even a substantial loss of ozone will not result in an increase in UV-B to the level experienced in summer or at lower latitudes. Observations like these have been used to argue that damage to the ozone layer hardly matters. People currently experience very different uv climates, the argument goes, so why all the fuss? Summarize *briefly* the points that you would make in response to an argument like this.

7 STRATEGIES FOR PROTECTING THE OZONE LAYER

STUDY COMMENT   In this final Section, we take a brief look at the stepwise process that has led to the phasing out of CFCs and several other synthetic halocarbons. You will not be expected to remember the historical details. They are included to give you a feel for the way the policy-making process has been – and continues to be – influenced by scientific research into the ozone issue.

7.1   Forging international agreement

Now that subsequent developments have vindicated the concern behind the Montreal Protocol, it is easy to forget the historical background to this seminal international agreement. Some of the key events in the decade or so that preceded it are collected in Box 7. In retrospect, this period can be viewed as an important testing ground for what has been termed the 'precautionary principle'. The arguments for and against this principle were at their most stark during the earliest phase of the CFC debate, which erupted in the USA in the mid-1970s.

In brief, those with an interest in maintaining the status quo argued that several major industries should not be jeopardized (with the attendent risks of unemployment and economic damage) on the strength of a *theoretical* prediction, unsupported by observations on the real atmosphere. They stressed the uncertainties in the calculations, and advocated a 'wait and see' approach. By contrast, those who argued for urgent action to control the release of CFCs stressed the unusual stability of the compounds: this, together with the natural variability of the ozone column, meant that monitoring the atmosphere could not provide an *early* warning of ozone loss. Rather, by the time a clear signal was detected, there would already be enough CFCs stored in the troposphere to keep the damage going for many decades, even if emissions were then halted at once. Just the circumstances we now find ourselves in.

BOX 7

A BRIEF CHRONOLOGY OF THE BACKGROUND TO THE MONTREAL PROTOCOL

- **1974**   Rowland and Molina publication identifies potential threat of CFC-linked ozone depletion.

- **1976**   First report from US National Academy of Sciences confirms continued release of CFCs as a 'legitimate cause for concern'.

- **1977**   United Nations Environment Programme (UNEP) recommends 'World Plan of Action on the Ozone Layer' and establishes coordinating committee to issue reviews of current research.

- **1978**   USA bans use of CFCs in non-essential aerosols (followed by Canada, Norway and Sweden).

- **1980**   European Community (EC) reduces aerosol use of CFCs by 30% and enacts cap on production capacity.

- **1982**   UNEP convenes working group of legal and technical experts charged with drawing up a draft framework convention for protecting the ozone layer.

- **1985**   Vienna Convention for the Protection of the Ozone Layer adopted in March. It declared the signatories' determination 'to protect human health and the environment against adverse effects resulting from modifications to the ozone layer', and provided, *inter alia*, for the exchange of information and scientific data relevant to this determination.

- **1986**   UNEP and World Meteorological Organization (WMO, another UN agency) publish *Atmospheric Ozone 1985*. Then the most comprehensive assessment ever undertaken, this report was a conscious effort 'to provide governments around the world with the best scientific information currently available on whether human activities represent a substantial threat to the ozone layer'. Its production involved around 150 scientists from various nations.

- **1986–1987**   Negotiations for the Montreal Protocol on Substances that Deplete the Ozone Layer, signed in September 1987.

As it turned out, unilateral action by the USA in 1978 (Box 7) was an important step towards acceptance of the precautionary principle by the wider international community. But it was a long and difficult process. At the outset, international industry was strongly opposed to regulatory action. Many of the governments whose cooperation was needed were equally hostile. And further research was providing substance to those who argued that the drive for international controls was premature – or unnecessary: recall that the early 1980s saw a reassuring downward trend in model forecasts of long-term ozone depletion (Figure 22 in Section 4.3).

Within the scientific community, however, there was an emerging consensus that the long-term dangers did warrant some sort of precautionary action. Consolidated through a succession of national and international assessments of the 'current state of knowledge', that consensus became an important driving force behind ozone policy. By taking a shared responsibility for the implications of their findings, scientists were drawn into an ongoing dialogue with political and economic decision-makers. That collaboration was a vital factor in the negotiations that eventually produced the **Montreal Protocol**, and the **Vienna Convention** from which it was born.

Signed by the EC and 20 nations in March 1985, this Convention specified no detailed targets or regulations concerning CFC emissions. But it did set an extraordinary precedent: it was the first time that nations agreed *in principle* to tackle a global environmental problem *before* its effects were manifest. News of the ozone hole broke two months later – too late to be included in the major UNEP/WMO study mentioned in Box 7. And it was that study – along with assessments of the technical and economic feasibility of phasing out the most damaging halocarbons – that formed the input to the original Montreal Protocol. As Richard Benedick (then head of the US delegation) has sought to stress:

> The ozone hole did not ... provide any clear signal for policy-makers at that time. Scientists in 1986 and 1987 were far from certain that CFCs were involved in Antarctica ... the phenomenon was contrary to known theory and did not conform to the global model predictions of gradual and pervasive long-term depletion. ... In achieving the treaty, consensus was forged and decisions were made on a balancing of probabilities. And the risks of waiting for more complete evidence were finally deemed to be too great.
>
> (Source: R. E. Benedick, 1991, *Ozone Diplomacy: New Directions in Safeguarding the Planet*, Harvard University Press)

The controls agreed at Montreal covered a group of fully halogenated compounds – five CFCs and three halons (*brominated* compounds: recall Exercise 1 in Section 3.6). The overall effect would have been to freeze halon consumption at 1986 levels in 1992, and to cut that of the CFCs to 50% by the end of the century. This hard-won agreement was signed on 16 September 1987. Just two weeks later the first results started coming in from Punta Areñas.

7.2 Science and the review process: the Montreal Protocol in evolution

An important element in the Protocol is Article 6, which provides for a periodic review of its control measures 'on the basis of available scientific, environmental, technical and economic information'. Against the backdrop of the disturbing scientific findings discussed in Section 5, two such reviews had already taken place at the time of writing – in London in June 1990, and in Copenhagen in November 1992. The London review mandated a complete phase-out – not only of the original CFCs and halons, but also of carbon tetrachloride and methyl chloroform. Two years later, parties to the Protocol agreed to bring forward the phase-out dates for all of these compounds. They also greatly expanded the scope of the treaty, enacting controls on a range of other partially halogenated compounds as well. The amended terms agreed in Copenhagen (and in force since 1994) are summarized in Table 6. Pertinent information on a selection of the controlled substances is collected in Table 7.

Table 6 Reductions in consumption of halocarbons according to the provisions of the Copenhagen amendment to the Montreal Protocol.[a]

Product	Action
Substances controlled at Montreal[b]	
chlorofluorocarbons (CFCs)	phase out by 1996, with 75% cut by 1994
e.g. CFC-11, CFC-12, CFC-113	
halons (Hs) e.g. H-1301	phase out by 1994
Substances controlled since Montreal[c]	
carbon tetrachloride (CT)	phase out by 1996, with 85% cut by 1995
methyl chloroform (MC)	phase out by 1996, with 50% cut by 1994
hydrochlorofluorocarbons (HCFCs)	cap consumption in 1996, and phase out by 2030
e.g. HCFC-22, HCFC-141b, HCFC-142b	
methyl bromide (MB)	freeze at 1991 levels in 1995, with subsequent cuts under review

[a] Cuts refer to levels of consumption in (b) 1986 and (c) 1989 (except for methyl bromide). The Protocol explicitly exempts the amounts used as chemical feedstock, and the amounts captured for recycling and/or destruction from being counted as consumption within each group of compounds. Small 'essential-use' exemptions to the 1996 CFC phase-out target were agreed in 1994, mostly for aerosol propellants in medical inhalants.

Table 7 A selection of the substances controlled under the evolving provisions of the Montreal Protocol: characteristics and uses.

Compound	Ozone depletion potential	lifetime / yr	Mixing ratio[a] / p.p.t.v.	World consumption in 1986/10^3 tonnes	Past, current and and potential uses[b]
Chlorinated compounds					
CFC-11 ($CFCl_3$)	1.0	65	255	411	A, PF, R, S
CFC-12 (CF_2Cl_2)	1.0	130	470	487	A, AC, PF, R
CFC-113 (CCl_2FCClF_2)	0.8	90	70	182	A, R, S
CT (CCl_4)	1.1	50	107	1 116	CF, S
MC (CH_3CCl_3)	0.15	6	160	609	A, Ad, S
HCFC-22 (CHF_2Cl)	0.05	15	103	140	A, AC, PF, R
HCFC-141b (CH_3CCl_2F)	0.1	10	–	c	A, PF, R, S
HCFC-142b (CH_3CClF_2)	0.06	20	–	c	A, AC, PF, R, S
Brominated compounds					
H-1301 (CF_3Br)	10	77	~2	11	FF
MB (CH_3Br)	0.7	~2	~11	~60	Ag

[a] In 1990 (as in Table 4 in Section 3.5). [b] Abbreviations for uses: A = aerosols; AC = air conditioning; Ad = adhesives; Ag = agriculture; CF = chemical feedstock; FF = firefighting; PF = plastic foams; R = refrigeration; S = solvents.
[c] Being developed as a CFC substitute (see Section 7.2.1).

Concentrate on the chlorinated compounds for now, and on the column headed **ozone depletion potential** (**ODP**) in Table 7. This index was introduced into the original Protocol as a guideline for estimating a compound's *potential* for inflicting damage to the ozone layer, with CFC-11 being used as an arbitrary standard. Formally, it is defined as follows:

$$\text{ODP (for compound X)} = \frac{\text{ozone depletion due to X}}{\text{ozone depletion due to CFC-11}}$$

where the ozone depletion is the model-calculated change in column ozone, *when the compounds have reached their steady-state atmospheric concentrations*, following prolonged release at the same constant rate.

Although the detailed mechanism whereby an individual halocarbon releases its load of chlorine in the stratosphere influences the calculated ODP value – as does the number of Cl atoms per molecule – the single most important factor is the compound's atmospheric lifetime. For a given uniform release rate, the longer a compound's lifetime, the higher its eventual steady-state concentration (Box 8). From this perspective, the long-lived *fully* halogenated compounds (a category that includes CCl_4, note) represent a greater potential threat to stratospheric ozone than their shorter-lived *partially* halogenated counterparts – a message evident in the ODP values recorded in Table 7.

But an ODP defined at steady state can be a potentially misleading parameter when considering the *short-term* impact of halocarbon emissions. As far as chlorinated compounds are concerned, this issue came to the fore during the London review: it warrants closer inspection.

BOX 8

ATMOSPHERIC LIFETIMES AND CONCENTRATION GROWTH

For simplicity, suppose that over a prolonged period a halocarbon is released to the atmosphere at a *constant* annual rate (p, say). Under these circumstances, the rate at which its atmospheric concentration (c) changes with time is given by:

dc/dt = release rate – loss rate

$\qquad = p - (c/\tau)$

where τ is the compound's atmospheric lifetime (as defined in Box 1). The compound will go on accumulating in the atmosphere until its loss rate matches the release rate. At this point, $dc/dt = 0$, and the concentration levels off at its *steady-state* value (c_{ss}), where

$\qquad c_{ss} = p\tau$

For a given constant release rate (p = 1 unit/year, say), a long-lived species eventually builds up to a much higher steady-state concentration than a short-lived one. Part (a) of the Figure illustrates the point for two hypothetical halocarbons, X and Y.

Notice that Y builds up to its final (albeit lower) steady-state concentration much more rapidly than does the longer-lived compound X. As a result, it also takes a long time for the *relative* concentrations of X and Y to approach the steady-state ratio determined by their respective lifetimes, as shown in part (b) of the Figure. Early on, the relative concentration of Y is much higher than this steady-state ratio would suggest.

This is why ODP values defined relative to the steady-state impact of a long-lived compound (i.e. CFC-11) can seriously underestimate the near-term effect of releasing short-lived species. Early on, the relative effect of the short-lived species Y would be higher than its ODP value might suggest because its potential steady-state impact on ozone is realized within some 20–30 years, whereas the effect from the long-lived species X is then well below its steady-state value.

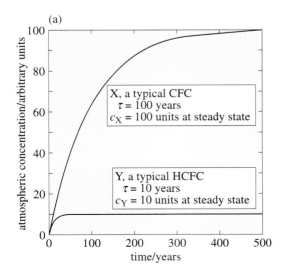

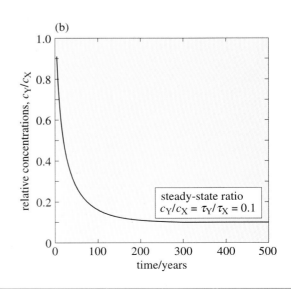

7.2.1 Restoring the ozone layer: the 'chlorine-loading' methodology

Acknowledging the failings of existing computer models, the scientific assessment for the 1990 review did not attempt to forecast future ozone losses in detail. Instead, it took the central problem to be the bleak prognosis for Antarctic ozone – and the uncertain, but possibly serious, implications for the rest of the planet – unless the *total* chlorine loading of the atmosphere is first stabilized, and then progressively reduced toward the level (estimated at some 2 p.p.b.v.) associated with the first appearance of the ozone hole. Calculated on a year-by-year basis, projections of future chlorine loadings were used to identify the control strategies needed to achieve this objective within the foreseeable future. Such projections include the contributions made by *all* chlorinated compounds – long- and short-lived, high and low ODP alike. Representative examples from the London review are collected in Figure 35.

The take-home message was clear. By itself, an early CFC phase-out would have little immediate effect on the upward trend in atmospheric chlorine: compare curves A and B. To get the recovery process underway would require drastic short-term action to curb the chlorine loading contributed by other compounds as well. With its long lifetime and high ODP (Table 7), carbon tetrachloride (CT) was an obvious target: like the CFCs, halting emissions of CT as soon as possible would begin the slow process of purging the compound from the atmosphere (curve C in Figure 35). The focus on methyl chloroform (MC) is more telling, reflecting the shift in emphasis away from the ODP concept. Taken alone, its low ODP value would rank MC as far less damaging than the CFCs (Table 7 again). But release enough of a relatively 'safe' species like this, and it can still make a significant contribution to the *total* chlorine loading: MC accounted for some 16% of the total in 1990 – more than double that due to CFC-113. On the other hand, the contribution from this short-lived species decays away rapidly once emissions cease (curve D in Figure 35).

It is clear from Figure 35 that the controls eventually agreed in 1990 would have achieved a striking reduction in atmospheric chlorine (curve D) – compared, that is, with the projected impact of the original Protocol (curve A). But even so, the chlorine loading was still predicted to be around 3 p.p.b.v. by the end of the next century. Yet stronger measures were needed to bring this down toward the desired 2 p.p.b.v. 'threshold'. That brought the spotlight to bear on the **hydrochlorofluorocarbons (HCFCs)**: one typical example, HCFC-22 (CHF_2Cl), is already a major refrigerant.

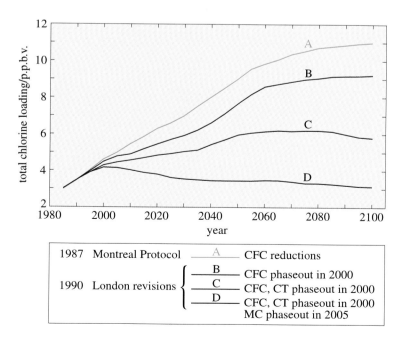

Figure 35 Projections of future chlorine loadings from natural methyl chloride, CFCs, carbon tetrachloride (CT), methyl chloroform (MC) and HCFCs arising from the different control strategies specified in the key. All scenarios assume that HCFCs (represented by HCFC-22) replace 30% of CFC reductions and are not subject to future controls.

As a glance back at Table 7 will confirm, HCFCs fall in the same general category as methyl chloroform: they have much shorter lifetimes and lower ODP values than the CFCs. These desirable characteristics encouraged the chemical industry to target HCFCs as among the most suitable alternatives to CFCs. By 1990, the industry had already invested huge amounts in a crash programme of research aimed at the development of new HCFCs, compounds that could be tailored to have physical properties closely matched to those of the original CFCs. Assessments at the time indicated that HCFCs were the only feasible alternatives in sight in certain critical sectors, especially in refrigeration and air conditioning. In short, these compounds looked set to capture around 30% of the existing and future market for CFCs – whence the assumption along these lines incorporated in the projections in Figure 35.

Moves to extend the Protocol to cover HCFCs proved unsuccessful in 1990. But the parties did approve a declaration that characterized these compounds as 'transitional substances', and called for their 'prudent and responsible' use prior to their phasing out between 2020 and 2040. The aim here was to signal to producers and consumers alike that they should not become too dependent on chlorinated substitutes. At the same time, it was recognized that premature phase-out dates could inhibit the additional investment needed to bring the new compounds to market – and that could, in turn, delay the rapid abandonment of old CFC technologies.

The intent behind the London declaration was formally adopted two years later in Copenhagen. As noted in Table 6, the agreed mechanism amounts to a cap on the consumption of HCFCs, imposed in 1996 and thereafter reduced stepwise, leading to a complete phase-out in 2030. Figure 36 highlights the extra chlorine loading that could arise if HCFCs are consumed up to the maximum amount allowed under the Copenhagen amendments (curve B), compared with that contributed by natural methyl chloride, the CFCs, carbon tetrachloride and methyl chloroform alone (curve A). On this basis, transitional reliance on HCFCs is expected to have little impact on the point at which the total chlorine loading falls to 2 p.p.b.v. – around the middle of the 21st century. That point is mainly determined by the agreed controls on the CFCs, CT and MC: as noted earlier, the phase-out schedules for these compounds were also tightened at Copenhagen. On the other hand, it is clear that using HCFCs as CFC substitutes does have the potential to add to the peak loading in the immediate future.

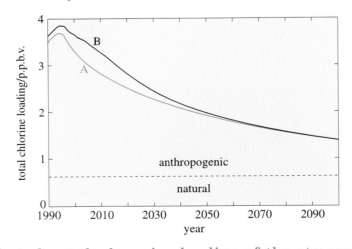

Figure 36   Projections of future chlorine loadings assuming global compliance with the 1992 amendments to the Montreal Protocol: (A) from natural methyl chloride, CFCs, carbon tetrachloride and methyl chloroform; (B) from all of the above, plus HCFCs (represented by HCFC-22), consumed at the maximum amount allowed.

7.2.2   What about the bromine loading of the atmosphere?

Just as the evolving provisions of the Montreal Protocol have mandated ever tighter controls over chlorinated compounds, so too another class of ozone-depleting substances, those containing bromine, have come under ever closer scrutiny.

■   Have another look at Table 7. Notice that H-1301 and CFC-11 have comparable atmospheric lifetimes. Yet the halon rates an ODP of 10, even though it carries only a single Br atom, whereas CFC-11 has three Cl atoms per molecule. Can you suggest why?

■   The underlying reason was foreshadowed in the answer to Exercise 1 (Section 3.6). In brief, bromine carried up into the stratosphere is more efficient at destroying ozone than is chlorine (by a factor of around 40, according to one recent estimate) mainly because relatively more of it is likely to be in the active forms Br· and BrO·.

Estimates like this explain why the long-lived halons were included in the original Protocol. Since then, the switch to a 'total loading' methodology has drawn attention to the presence in the atmosphere of a large number of brominated compounds. There has never been a systematic attempt to monitor the complete set of compounds on a global scale, and there remain large uncertainties about their various natural and anthropogenic sources. Nevertheless, reducing the bromine loading contributed by human activities is now seen as an important element in the overall strategy for restoring the ozone layer. The Copenhagen amendment (Table 6) took an important step in that direction, by including methyl bromide among the controlled substances. Here, the rationale is a familiar one. Large amounts of methyl bromide are manufactured for agricultural purposes: it is a widely used soil fumigant, for example. Curbing emissions of this short-lived ($\tau \cong 2$ years) species would achieve a rapid reduction in the amount of Br being carried up into the stratosphere.

7.3 Looking to the future

The Montreal Protocol was designed to encourage wide participation. At the time, the industrialized nations accounted for some 88% of the total global consumption of CFCs: per capita usage was over 20 times higher than in the developing world. Nevertheless, it was recognized that meeting the aspirations for higher living standards in developing countries, especially in the area of commercial and industrial refrigeration, could lead to a significant increase in their per capita consumption of CFCs. And with over 75% of the world's population, that could undermine the North's efforts to protect the ozone layer.

The original Protocol sought to encourage the participation of developing countries by exempting them from controls for a 10-year 'grace period' – even as significant cutbacks took effect in the industrialized world. On top of this, the London review saw agreement to a financial mechanism under which developed countries will meet the incremental costs that developing countries incur in complying with the Protocol. This is intended to cover all aspects of the process of technology transfer, including access to the necessary technical expertise, patents, training and so on, as well as the direct costs of converting existing production facilities or of establishing new ones, or of simply importing CFC substitutes. The heart of the mechanism is a special Multilateral Fund, set at an interim level of 240 million US dollars for the first three years. This was put on a permanent basis at Copenhagen, with funding of 510 million US dollars for the period 1994–1996.

The establishment of a permanent Multilateral Fund brought the crucial participation of China and India, the world's two most populous nations. By late 1994 there were 148 parties to the Protocol: all of the industrialized countries had passed national – or in the case of the European Union, community – laws or regulations to implement the phase-outs, with many moving ahead of international targets in at least some way.

Details apart, the point to note is that the projections in Figure 36 assume that the revised (1992) treaty obligations (or even tighter controls) are followed to the letter, and that non-parties do *not* commence large-scale production of the controlled substances. Even under these circumstances, the world could still see a seasonal recurrence of the Antarctic ozone hole for decades to come. The chances that it could also see an Arctic ozone hole – and/or new record ozone lows elsewhere – are more difficult to assess. Given the unresolved issues highlighted in Section 5.3, it is recognized that modelling studies cannot offer detailed guidance on this front. The interactions and feedbacks noted in Section 6.2 only compound the problem by pointing to the importance of factors like the atmospheric concentration of CO_2, and the way that ozone depletion feeds back on stratospheric temperature.

Indeed, perhaps the most important general lesson to take from this Topic Study is that predicting how a complex natural system, such as the atmosphere, will respond to human interference is fraught with difficulties. Computer models are the best tools we have, but they can only ever be approximate representations of the 'real world' system. Their forecasts of what the future might be like are beset with uncertainties. On the other hand, the history of ozone research provides powerful support for the view that such uncertainties always cut two ways: there is no *a priori* reason to suppose that incompletely understood factors – or hitherto unsuspected mechanisms – will necessarily act to mitigate the perceived threat. They could, equally well, push the response of the real system beyond that indicated by model simulations.

As far as the future health of the ozone layer is concerned, all that can be said with certainty is that the *potential* exists for further attrition, especially during the period when levels of atmospheric chlorine and bromine are expected to go on rising – probably until sometime between 1998 and 2010, on current estimates.

On a more upbeat note, the story of 'ozone diplomacy' to date is an inspiring example of foresight in tackling a *global* environmental problem, and constructive cooperation between governments, industry and science. As a glance back at Figure 35 (curve A) will confirm, the regulatory regime put in place since Montreal has done much to avert the prospect of deeper ozone losses. Further, 1994 saw the first evidence that the Protocol is working: growth rates in the atmospheric concentrations of several controlled substances have begun to decline. Even so, researchers versed in the problems of *implementing* international regimes warn that there are several current and emerging difficulties. To quote from a recent article on the subject:

> First, controls on ozone depleting substances in industrial countries are now becoming stringent and costly for the first time, with predictable consequences: a political backlash against the protection effort, including attacks on the scientific consensus about ozone depletion; a black market [in CFC imports, needed to recharge existing equipment]; and increased suspicions of cheating or circumvention ... Second, implementing phaseout projects in developing countries has been much harder and slower than anticipated.
>
> (Source: Edward A. Parson and Owen Greene, *Environment*, 1995, Vol. 37, No. 2, pp. 16–20 and 35–43)

Ten years on from the Vienna Convention, the next review of the Protocol's core commitments is due to take place in Vienna in November 1995. Alongside the negotiation of new targets and timetables (likely to focus on the HCFCs and methyl bromide), parties to the Protocol are faced with the complex task of agreeing measures to promote the effective implementation of *existing* controls. It remains to be seen whether the international community can sustain its commitment to the ozone layer as progress becomes more difficult.

Postscript

In October 1995, it was announced that the Nobel prize for Chemistry had been awarded jointly to Paul Crutzen of the Max Planck Institute for Chemistry in Mainz, Mario Molina of the Massachusetts Institute of Technology and F. Sherwood Rowland of the University of California at Irvine, for their work in atmospheric chemistry. The citation from the Royal Swedish Academy of Sciences highlights in particular their role in identifying the threat of stratospheric ozone depletion, and in elucidating the mechanisms by which this occurs.

Born in the Netherlands, Paul Crutzen drew attention in 1970 to the catalytic role of NO_x in the natural stratosphere. The following year, Harold Johnston (an atmospheric chemist at the University of California at Berkeley) pointed out that the NO_x released by supersonic aircraft (dubbed supersonic transports or SSTs) that fly within the stratosphere might lead to significant ozone depletion. This work first brought to the fore the possibility that human activities could pose a threat to stratospheric ozone. The scientific effort put in motion by the debate about SSTs soon identified other potential modifiers of the ozone layer – including atomic chlorine. Together with James Lovelock's measurements of trace levels of CFCs throughout the global atmosphere (first reported in 1972), that set the scene for Rowland and Molina's initial warning about CFCs.

Paul Crutzen   Mario Molina   F. Sherwood Rowland

OBJECTIVES FOR TOPIC STUDY 1

Now that you have completed Topic Study 1, you should be able to do the following things:

1   Recognize valid definitions of, and use in a correct context the terms, concepts and principles printed in bold type in the text and collected in the following Table.

Term	Page No.
Antarctic ozone hole	38
atmospheric lifetime, τ	21, 24
catalytic cycle	17
catalytic families (BrO_x, ClO_x, NO_x and HO_x)	18
Chapman mechanism	12
chlorine loading	22
chlorofluorocarbons (CFCs)	22
column ozone	29
greenhouse gases	34
halons	26
hydrochlorofluorocarbons (HCFCs)	59
mixing ratio by volume	8
Montreal Protocol	56
number density	8
odd oxygen (O_x)	15
ozone depletion potential (ODP)	57
ozone layer	9
photodissociation (photolysis)	11
photodissociation coefficient, j	13
polar stratospheric clouds (PSCs)	39
polar vortex	39
reservoir molecule	20
source gases	21
stratosphere	7
tropopause	7
troposphere	7
Vienna Convention	56

2   Given appropriate information, estimate the number density of air, or of a specified atmospheric constituent, at a particular altitude. (SAQ 1)

3   Given appropriate information, estimate (and/or comment on the significance of) the atmospheric lifetime of a specified compound. (SAQs 6 and 7; Exercises 1 and 2)

4   Use the concept of odd oxygen in the kinetic analysis of a reaction scheme relevant to stratospheric chemistry, or discuss the implications of the results of such an analysis. (SAQs 2, 3, 4 and 8; Exercise 1)

5   Demonstrate an understanding of:

(a)   the main features of the natural system that maintains the stratospheric ozone layer, including both chemical and dynamic factors;

(b)   the mechanism that is believed to be responsible for the Antarctic ozone hole;

(c)   the reasons why ozone depletion is (currently) less severe over the Arctic than over Antarctica;

(d)   the processes that may contribute to ozone depletion at middle latitudes.

(SAQs 5 and 9; Exercises 1 and 2)

6   Drawing on the understanding referred to in Objective 5, comment on:

(a)   the human activities and natural factors that are likely to influence the future evolution of the ozone layer;

(b)   the limitations of the computer models that have been used to predict CFC-linked ozone depletion over the past 20 years.

(SAQ 11; Exercise 2)

7   Demonstrate an awareness of the possible consequences of ozone layer depletion. (SAQ 12)

SAQ ANSWERS AND COMMENTS

SAQ I (Objectives I and 2)

(a) $[M] = 8.3 \times 10^{17}\,cm^{-3}$; (b) $[O_2] = 1.7 \times 10^{17}\,cm^{-3}$; (c) $p_{O_2} = 5.0\,mbar = 500\,Pa$.

(a)   The number density $[M] = N/V$ is calculated from the ideal gas equation as follows:

$pV = nRT = (N/L)RT$ (where L is the Avogadro constant), so

$[M] = N/V = pL/RT$

At 25 km, $T = 219\,K$ and $p = 25.0\,mbar = 25.0 \times 10^2\,Pa = 25.0 \times 10^2\,J\,m^{-3}$. So

$$[M] = \frac{(25.0 \times 10^2\ J\,m^{-3}) \times (6.022 \times 10^{23}\ mol^{-1})}{(8.314\ J\,K^{-1}\,mol^{-1}) \times (219\ K)}$$
$$= 8.27 \times 10^{23}\,m^{-3} = 8.3 \times 10^{23}\,(10^2\,cm)^{-3}$$
$$= 8.3 \times 10^{17}\,cm^{-3}$$

(b) According to Table 1, the mixing ratio (i.e. fractional abundance) of O_2 is ~0.2, so

$[O_2] = 0.2[M] = 1.7 \times 10^{17}\,cm^{-3}$

(c) According to the definition of partial pressure (Block 1),

$p_{O_2} = n_{O_2}RT/V = (N_{O_2}/V)(RT/L) = [O_2]RT/L$

Thus, p_{O_2} could be calculated from the value of $[O_2]$ in part (b). But there's a simpler way. Since the *total* pressure at a given altitude is just $p = [M]RT/L$, and since $[O_2] = 0.2[M]$, it follows that

$$p_{O_2} = 0.2p = 0.2 \times 25.0 \, \text{mbar}$$

$$= 5.0 \, \text{mbar}$$

$$= 500 \, \text{Pa}$$

SAQ 2 (Objectives 1 and 4)

The short answer to the second question is 'yes'. According to the observational data recorded in Figure 10, $[O]$ rises and $[O_3]$ falls with increasing altitude. Thus, the ratio $[O]/[O_3]$ increases with increasing altitude, a trend that is consistent with the predictions of equation 11:

$$\frac{[O]}{[O_3]} = \frac{j_3}{k_2[O_2][M]}$$

With increasing altitude, j_3 increases (slightly, Figure 8), but each of the terms in the denominator decreases; recall that temperature rises with altitude in the stratosphere, and the temperature-dependence of k_2 is such that its value falls (slightly) with increasing temperature. Working together, these effects cause the ratio $[O]/[O_3]$ to increase with altitude – as observed.

SAQ 3 (Objective 4)

(a) With the assumptions in the question, equations 7 and 8 become:

$$\frac{d[O]}{dt} = 2j_1[O_2] - k_2[O][O_2][M] + j_3[O_3] - k_4[O][O_3] = 0$$

$$\frac{d[O_3]}{dt} = k_2[O][O_2][M] - j_3[O_3] - k_4[O][O_3] = 0$$

Adding these equations together does indeed produce the required expression.

(b) Substituting $k_4[O][O_3] = j_1[O_2]$ into the expression for $d[O_3]/dt$ in part (a) produces

$$\frac{d[O_3]}{dt} = k_2[O][O_2][M] - j_3[O_3] - j_1[O_2] = 0 \text{ (as required)}$$

This expression can be reorganized to give:

$$k_2[O][O_2][M] = j_3[O_3] + j_1[O_2]$$

whereas the analysis in the text (i.e. treating steps 2 and 3 as a rapidly established equilibrium) gave:

$$k_2[O][O_2][M] = j_3[O_3] \tag{10}$$

Comparing these two expressions suggests that the second one involves the implicit assumption that $j_1[O_2] \ll j_3[O_3]$. This turns out to be a reasonable approximation under stratospheric conditions – the most important factor here being the relative values of j_1 and j_3 (recall Figure 8).

SAQ 4 (Objective 4)

(a) According to the extended Chapman mechanism in the text, steps 1, 4, 5 and 6 control the rate of change of $[O_x]$. Taking account of the change in odd oxygen in each step gives

$$\frac{d[O_x]}{dt} = 2j_1[O_2] - 2k_4[O][O_3] - k_5[X][O_3] - k_6[XO][O]$$

[If you are unsure about this, check that you get the same result by writing individual expressions for $d[O]/dt$ and $d[O_3]/dt$, and then adding these expressions together.]

(b) Applying the steady-state approximation to the reactive ('catalytic') species X leads to the following expression:

$$\frac{d[X]}{dt} = -k_5[X][O_3] + k_6[XO][O] = 0$$

It then follows that:

$$k_5[X][O_3] = k_6[XO][O]$$

Substituting for $k_5[X][O_3]$ in the expression for $d[O_x]/dt$ in part (a) leads to equation 16, as

$$\frac{d[O_x]}{dt} = 2j_1[O_2] - 2k_4[O][O_3] - 2k_6[XO][O]$$

$$= 2j_1[O_2] - 2\{k_4[O][O_3] + k_6[XO][O]\} \qquad (16)$$

SAQ 5 (Objectives 1 and 5)

The completed version of Figure 14 is shown here as Figure 37.

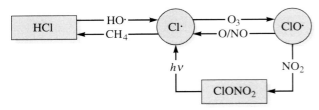

Figure 37   Completed version of Figure 14.

The links a = O_3 and b = O represent the catalytic cycle in equations 14 and 15, with X = Cl·. That is,

$$Cl· + O_3 \longrightarrow ClO· + O_2$$

$$ClO· + O \longrightarrow Cl· + O_2$$

I hope you spotted that the link c = NO can be deduced from Figure 12: it represents one of the reactions that interconverts the NO_x species,

$$ClO· + NO \longrightarrow Cl· + NO_2$$

Similarly, the link d = NO_2 represents the formation of the reservoir species $ClONO_2$ (again included in Figure 12), although strictly this should include a third body (for the reasons mentioned in Section 3.2), as:

$$ClO· + NO_2 + M \longrightarrow ClONO_2 + M$$

Finally, the links e = CH_4 and f = HO· represent the formation and breakdown of HCl – the main chlorine reservoir in the stratosphere (equations 17 and 18 in Section 3.4).

SAQ 6 (Objectives 1 and 3)

According to the definition of reaction half-life in Block 2, it would take one half-life to reduce the atmospheric concentration to a half, and two half-lives to reduce it to a quarter. Taking information from Table 4:

For CFC-11, $t_\frac{1}{2} = 0.693 \times 65 = 45$ yr, so the answers are (a) 45 years, and (b) 90 years.

For CFC-12, the atmospheric lifetime (τ) is double that for CFC-11, giving (a) 90 years, and (b) 180 years.

SAQ 7 (Objectives 1 and 3)

The two compounds in question (CH_3CCl_3 and CHF_2Cl) are not *fully* halogenated hydrocarbons. Given the discussion in Section 3.5 (and Box 1, in particular), plausible loss mechanisms for these compounds are chemical reactions within the *lower* atmosphere, triggered off by hydrogen-abstraction by the HO· radical, as:

$$CCl_3CH_3 + HO\cdot \longrightarrow CCl_3CH_2\cdot + H_2O$$

and

$$CHF_2Cl + HO\cdot \longrightarrow CF_2Cl\cdot + H_2O$$

In each case, the molecular fragment formed is itself a free radical, so it should engage in further reactions: and it does. The important general point is that the removal of these compounds is not dependent on slow transport up into the stratosphere. And this shortens their overall atmospheric lifetimes, just as it does for methane and methyl chloride.

SAQ 8 (Objective 4)

According to the expression in the question, *at a given altitude* (i.e. given value of j_1) the response to a change in temperature is determined by the temperature-dependence of the rate constants for the loss of odd oxygen – both directly, k_4 (via $O + O_3 \longrightarrow 2O_2$) and catalytically, k_6 (via $XO + O \longrightarrow X + O_2$). Reference to the activation energies collected in Table 2 reveals that the catalytic processes show little temperature-dependence: the predominant factor is the relatively strong temperature-dependence of k_4, with destruction of odd oxygen via the direct route proceeding *more slowly at lower temperatures*. On this basis, then, stratospheric cooling would be expected to shift the balance between creation and loss toward higher ozone concentrations.

[In practice, the 1-D model used to generate the results in Figure 23d incorporates a far more complete description of the ozone chemistry in Figure 18, including the workings of the storage system for the various catalytic species, and the way this depends on temperature. Such a model produces predictions that are broadly consistent with the simple arguments above because most of the temperature-dependence does indeed come from the Chapman oxygen-only reaction.]

SAQ 9 (Objective 5)

(a) This prediction is linked to the influence of the holding cycles that involve the two stratospheric reservoirs of active chlorine. The peak ozone loss occurs at altitudes with the highest mixing ratio of 'free' $ClO\cdot$. Lower in the stratosphere, models predict that most of the available chlorine should be locked up as HCl and $ClONO_2$ – a pattern that is also evident in the observational data included in Box 3.

(b) The key point is that Figure 23a records the model results as a percentage change in the *local* ozone concentration. In practice, the latter peaks at around 25 km (Figure 3), an altitude at which the percentage change is predicted to be close to zero. Larger percentage changes at higher altitudes (where ozone is less abundant) have a relatively small effect on the *total* ozone column.

SAQ 10 (revision)

The *net* effect of the catalytic cycle is the process:

$$2O_3 \longrightarrow 3O_2$$

Thus, the overall rate of reaction, J, is given by:

$$J = -\frac{1}{2}\frac{d[O_3]}{dt} = \frac{1}{3}\frac{d[O_2]}{dt}$$

If formation of the $ClO\cdot$ dimer (equation 27) is the rate-limiting step, then the overall rate of reaction can be set equal to the rate of this step, that is:

$$J = k_{27}[ClO\cdot]^2[M]$$

So

$$-\frac{d[O_3]}{dt} = 2k_{27}[ClO\cdot]^2[M]$$

On this basis, the rate of ozone loss would be expected to increase as the *square* of the concentration of the available ClO·. The latter could have increased by as much as $(2.5/1.2)^2$ – that is, more than four-fold – between 1970 and 1980. By 1990, atmospheric chlorine levels were roughly three times the 1970 value, implying a possible nine-fold increase in the rate of ozone loss. This quadratic dependence of the rate of the destruction chain on the chlorine loading of the atmosphere *is* thought to be a major factor in the sudden appearance of the ozone hole.

SAQ 11 (Objective 6)

The key point here is that the modelling studies referred to in the question did *not* include the influence of the heterogeneous processes discussed in Section 5. As noted earlier (see the answer to SAQ 8), *according to gas-phase chemistry alone*, ozone destruction proceeds more slowly at lower temperatures. Thus, the expected stratospheric cooling due to enhanced concentrations of CO_2 *may* reduce ozone loss, partially offsetting that from chlorine-catalysed ozone depletion. This temperature feedback remains of potential importance in the middle to upper stratosphere.

By contrast, a cooling of the *lower* stratosphere could lead to enhanced ozone loss. It could, for example, increase the abundance of PSCs and/or the time for which they persist, leading to a greater activation of reactive chlorine in polar regions. It might thereby produce conditions that would trigger the appearance of an Arctic ozone hole, or further increase the areal extent of the Antarctic hole, or enhance the influence of polar processes on ozone at lower latitudes. A general cooling of the lower stratosphere could also increase the possibility of more extensive chlorine-activation on sulfate aerosols.

SAQ 12 (Objective 7)

The important points can be summarized as follows:

- At any given location, ozone depletion would raise the mean UV-B flux about which natural fluctuations occur. This would increase the accumulated dose: it could also expose organisms to levels of UV-B higher than ever before experienced – a situation that is already occurring in springtime in the oceans around Antarctica. Further, the expected shift in spectral distribution would increase the intensity of the shorter, more 'biologically effective' wavelengths in the UV-B band.

- Even under an intact ozone shield, UV-B has harmful effects on human health that worsen with accumulated dose, the best documented example being the incidence of non-melanoma skin cancer. UV-B exposure is also believed to be a factor in the more deadly melanoma skin cancer and in the development of cataracts, and it can weaken the immune system.

- Changes in personal behaviour would help to mitigate the direct health-related effects of moderate ozone depletion. But to focus solely on this issue ignores all the other possible consequences of enhanced UV-B – for food supplies, and for terrestrial and aquatic ecosystems in general. Possible disturbance to the climate adds to the highly uncertain, but potentially serious, risks of sustained ozone depletion.

ANSWERS TO EXERCISES

Exercise 1 (Objectives 1, 3, 4 and 5)

(a) The simplest catalytic cycle is the one in equations 14 and 15 (Section 3.3), with X = Br·, that is:

$$Br· + O_3 \longrightarrow BrO· + O_2 \tag{32}$$
$$BrO· + O \longrightarrow Br· + O_2 \tag{33}$$

net: $O + O_3 \longrightarrow 2O_2$

Like the other X $\longrightarrow$ XO $\longrightarrow$ X catalytic cycles considered in Section 3.3, the potential importance of this loss mechanism can be assessed in terms of its contribution to $d[O_x]/dt$ via equation 16:

$$d[O_x]/dt = 2j_1[O_2] - 2\{k_4[O][O_3] + k_6[XO][O]\} \tag{16}$$

where, as before, k_6 refers to the second step in the catalytic cycle – reaction 33 in this case. Once again, then, you would need two kinds of information:

- The value of the rate constant for reaction 33 (k_{33}) under stratospheric conditions – the source being kinetic measurements, and hence values of k_{33} as a function of temperature. [In practice, it turns out that k_{33} is independent of temperature (i.e. the reaction has zero activation energy, so the term $e^{-E_a/RT}$ in the Arrhenius equation becomes $e^0 = 1$): the value $k_{33} = 3.0 \times 10^{-11}$ cm^3 s^{-1} is comparable with those for the other catalytic reactions collected in Table 2.]

- The concentration (as a function of altitude) of BrO·. One important factor here is obviously the 'supply' of bromine atoms, via the breakdown in the stratosphere of source gases (such as CF_3Br) that are transported up from the lower atmosphere: a full assessment would need to incorporate information about the emissions (and atmospheric lifetimes) of *all* possible source gases – both natural and anthropogenic. More on that in Section 7. But given the discussion in Section 3.4, there is also another factor here: this is the subject of part (b).

(b) The following is *not* a 'model answer': rather, it simply lists the key points that you should include in a discussion of the proposition posed in the question.

1   To begin with, a brief resumé of the points in part (a) would be in order. The thing to stress is that the information in Table 5 allows you to comment on the extent to which 'active' bromine radicals (Br· and BrO·) are likely to become locked up as the plausible reservoir species HBr and $BrONO_2$, compared with the situation for chlorine.

2   Under stratospheric conditions, the formation of HCl diverts a significant fraction (some 70%) of the 'available' chlorine. The formation of HBr (via the analogous reaction with CH_4, the second column in Table 5) is likely to be far less effective in diverting active bromine, on both thermodynamic and kinetic grounds.

- *Thermodynamics*: According to the relation $\Delta G_m^{\ominus} = -RT \ln K^{\ominus}$ (Block 1), the more negative the value of $\Delta G_m^{\ominus}$ (at a particular temperature), the larger is $K^{\ominus}$ and the more thermodynamically favourable is the reaction. Further, $\Delta G_m^{\ominus}$ under stratospheric conditions (i.e. $T = 220$ K) can be calculated from the relation $\Delta G_m^{\ominus} = \Delta H_m^{\ominus} - T\Delta S_m^{\ominus}$. Here, the value of $\Delta S_m^{\ominus}$ is likely to be small (no change in the number of gaseous molecules) and similar in magnitude for the two reactions. Both reactions are endothermic ($\Delta H_m^{\ominus} > 0$), but the larger positive value for the bromine reaction suggests that this process is likely to be much less thermodynamically favourable at a given temperature.

Kinetics: For an *endothermic* elementary reaction, the value of $\Delta H_m^{\ominus}$ represents a *lower* limit to the size of the activation energy, E_a (see Figure 38). Thus, the energy barrier to reaction is likely to be much higher for the Br reaction. (Refer back to the discussion of potential energy surfaces in Block 2 if you missed this point.)

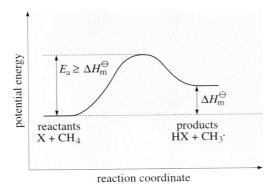

Figure 38   A schematic potential energy profile for an endothermic elementary reaction.

3   The information in the third and fourth columns of Table 5 relates to the formation and breakdown (via photolysis) of bromine nitrate ($BrONO_2$), which is another possible reservoir of active bromine. Under stratospheric conditions, the rate constant for the formation of this species is an order of magnitude larger than that for the analogous chlorine species. But the data provided indicate that it also has a much larger photodissociation constant (j), and hence a much *shorter* (photolytic) lifetime—in the mid-stratosphere (25 km) at least. I didn't expect you to do this, but it's worth noting that these lifetimes can be estimated as:

$ClONO_2$: $\tau = 1/j = (1/7.4 \times 10^{-5})\,s \cong 4\,hours$

$BrONO_2$: $\tau = 1/j = (1/1.6 \times 10^{-3})\,s \cong 10\,minutes$

In short, active forms of bromine are again favoured.

4   In conclusion: *According to the information provided*, natural 'checks' on the ozone-destroying potential of bromine are less effective than they are for chlorine: relatively more of the bromine in the stratosphere is likely to be in active (Br· and BrO·) ozone-destroying forms. [In principle, there *could* be other reservoir species for bromine: in practice, none has yet been identified.]

Exercise 2 (Objectives 1, 3, 5 and 6)

There is no one 'correct' way of tackling an open-ended question like this. But as a guide, the following list reflects *my* judgement of the points that should feature in the proposed article. In practice, an important constraint would be the specified word limit: you would need to be selective, and to make your points clearly and succinctly. To that end, careful use of *annotated* diagrams can be a good general strategy: I have included reference to a few illustrative examples.

1   The 'punchy' opening paragraph should stress that ozone depletion is now manifestly evident – most dramatically over Antarctica in springtime (Figure 25 would make the point, plus allow you to 'place' the ozone layer in the stratosphere; a couple of TOMS 'maps' would be more dramatic, if you had colour!), but also (albeit more gradually) at latitudes that take in large tracts of North America, Europe and Asia.

2   The bulk of the article should cover the following points:

- Ozone amounts are maintained by a balance between production and loss. Something like Figure 7 would help here, allowing you to then point out how trace species (focus on Cl) reacting in a cyclic manner (Cl· $\longrightarrow$ ClO· $\longrightarrow$ Cl·) help to keep the ozone budget balanced at its 'normal' level.

Hence the original concern about CFCs. Released at the surface, these unusually stable compounds are stored up in the lower atmosphere, whence they are slowly cycled up into the harsh photochemical environment in the stratosphere. Their breakdown adds to the background level of chlorine, threatening to disturb the natural balance and cause a shift to a new steady state that sustains less ozone.

There is good evidence that chlorine chemistry *is* responsible for the seasonal recurrence of the Antarctic ozone hole, but the mechanism involved is more complex. In brief, the unique meteorology of the Antarctic stratosphere in winter (an intensely cold mass of air, well laced with PSCs, 'contained' by a strong polar vortex) sets up conditions that effectively remove the normal checks on chlorine-catalysed ozone loss (the reservoir molecules). In my view, the most effective way to get the essential chemistry across would be via an annotated diagram like the one in Plate 5: a figure like this was used in an article by Owen B. Toon and Richard P. Turco (two American atmospheric scientists long engaged in ozone research) that appeared in *Scientific American* in June 1991.

The requirements for severe ozone loss (an ozone hole) are very cold temperatures (below about $-78\,°C$), for PSCs to 'activate' the chlorine, and for this activated air to be exposed to sunlight (recall that the catalytic cycle in Figure 27 depends on the formation *and photolysis* of the ClO· dimer). These conditions are always met during the Antarctic spring. But in the Arctic stratosphere, winters are usually warmer and shorter: PSCs are less abundant, and the polar vortex is weaker and is generally disrupted by warmer air pushing in from lower latitudes *before* the polar cap is fully sunlit. That is what happened in February 1992.

However – and this would be a crucial point to stress – the Arctic stratosphere is 'pre-conditioned' for chlorine-catalysed ozone loss: wintertime concentrations of ClO· can be as high as any ever observed over Antarctica. That was the point evident in the ER-2 measurements in January 1992 (later confirmed by data form the satellite-borne MLS).

Even without an Arctic hole as such, air pre-conditioned in polar regions may contribute to the unexpectedly large springtime losses of ozone at northern mid-latitudes.

3   In my view, the following points are pertinent to a paragraph or two dealing with 'future prospects'.

The heavily populated northern hemisphere has been spared an Arctic ozone hole because the more disturbed wintertime circulation has thus far negated the effect of chlorine activation by heterogeneous chemistry on PSCs. Comment on the possible effect of a CO_2-induced cooling of the lower statosphere (SAQ 11) would be appropriate.

A *brief* reference to the ozone anomalies in the aftermath of major volcanic eruptions (whatever the precise mechanism involved), to make the point that future ozone levels will also be subject to unpredictable natural events.

More generally, there have been many surprises over the past 20-odd years: further surprises may yet be in store.

Comment on the very long atmospheric lifetimes of the CFCs would be vital. The total chlorine loading of the atmosphere is now roughly six times its natural level (Section 3.5). Once emissions cease, it will take many decades for the atmosphere to cleanse itself of its burden of CFCs (SAQ 6). Until chlorine levels fall below the threshold values achieved in the late 1970s, the *potential* for an Antarctic ozone hole to appear each year will remain.

ACKNOWLEDGEMENTS

Grateful acknowledgement is made to the following sources for permission to reproduce material in this Topic Study:

Text

Extract in Box 4 from *Stratospheric Ozone 1988*, UK Stratospheric Ozone Review Group, © Crown Copyright. Reproduced with the permission of the Controller of Her Majesty's Stationery Office.

Figures

Figures 1, 3, 12, 14, 16, 23 and 37: *Stratospheric Ozone 1987*, UK Stratospheric Ozone Review Group, © Crown Copyright. Reproduced with the permission of the Controller of Her Majesty's Stationery Office; *Figure 2* Wayne, R. P. (1991) *Chemistry of Atmospheres: an Introduction to the Chemistry of the Atmospheres of Earth, the Planets and their Satellites*, © Richard P. Wayne 1985, 1991, Clarendon Press, reproduced by permission of Oxford University Press; *Figure 4* From *Handbook of Geophysics and Space Environments* by Air Force Cambridge Research Laboratories, edited by Shea L. Valley, McGraw-Hill Book Company, 1965; *Figure 5* Thomas, L. and Bowman, M. R. (1969) 'Atmospheric penetration of ultraviolet and visible solar radiations during twilight periods', in *Journal of Atmospheric and Terrestrial Physics*, **31**, p. 1311, reproduced by permission of Pergamon Press Ltd; *Figure 8a* DeMore, W. B. *et al.* (1992) *Chemical Kinetics and Photochemical Data for Use in Stratospheric Modeling: Evaluation Number 10*, August 15 1992, National Aeronautics and Space Administration; *Figure 8b* Brasseur, G. and Solomon, S. (1986) *Aeronomy of the Middle Atmosphere*, 2nd edition, Copyright 1986 by Kluwer Academic Publishers; *Figure 9* Professor Julius Chang; *Figure 10* Hudson, R. (ed.) (1981) *The Stratosphere 1981*, World Meteorological Organization, Geneva; *Figure 13 and Box 3* Reprinted with permission from McElroy, M. B. and Salawitch, R. J. (1989) 'Changing composition of the global stratosphere', in *Science*, **243**, Copyright 1989 American Association for the Advancement of Science; *Figure 15* Makide, Y., Yokahata, A., Kubo, Y. and Tominaga, T. (1987) 'Atmospheric Concentration of Halocarbons in Japan in 1979–1986, in *Bulletin of the Chemical Society of Japan*, **60**, pp. 571–574; *Figure 17* Warr, D. C. (1990) 'The Path to Ozone Loss', in *New Scientist*, 27 October 1990, © IPC Magazines Ltd, World Press Network 1990; *Figure 18* From *Atmospheric Change: An Earth System Perspective*, by Graedel, T. E. and Crutzen, P. J., Copyright © 1993 by AT&T. All rights reserved, W. H. Freeman and Company. Used with permission; *Figure 19* Dütsch, H. U. (1974) in *Canadian Journal of Chemistry*, **52**, p. 1491, © National Research Council of Canada 1974; *Figure 21* Henderson-Sellers, A. and McGuffie, K. (1987) *A Climate Modelling Primer*, John Wiley and Sons Ltd. Reprinted by permission of John Wiley and Sons Ltd; *Figure 22* Waters, J. W. (1993) 'The Chlorine Threat to Stratospheric Ozone' in *Engineering & Science*, **LVI**(4), Summer 1993, © 1993 Alumni Association, California Institute of Technology; *Figure 24* Based on a personal communication from Jonathan Shanklin, British Antarctic Survey; *Figure 25* Hofmann, D. (1992) in *NOAA 1992 Southern Hemisphere Winter Summary: selected indicators of stratospheric climate, NOAA climate analysis*, National Oceanographic and Atmospheric Administration; *Figures 26 and 28 and Box 4*: *Stratospheric Ozone 1988*, UK Stratospheric Ozone Review Group - Second Report, © Crown Copyright. Reproduced with the permission of the Controller of Her Majesty's Stationery Office; *Figure 27 and Box 5* Hamill, P. and Toon, O. B. (1991) 'Polar stratospheric clouds and the ozone hole', in *Physics Today*, December 1991, Copyright © 1991 American Institute of Physics; *Figure 29* Stolarski, R. S. (1991) 'Total ozone trends deduced from Nimbus 7 TOMS data', in Stolarski, R. S., Bloomfield, P., McPeters, R. D. and Herman, J. R., *Geophysical Research Letters*, **18**(6), June 1991, © 1991 American Geophysical Union; *Figure 30* McCormick, M. P., Thomason, L. W. and Trepte, C. R. (1995) 'Atmospheric effects of the Mt Pinatubo eruption', Reprinted with permission

Plates